This book is dedicated to my wife and best friend, Kandace, for her love and support in everything I do. In addition I want to dedicate this to my three wonderful children: my daughters, Natasha and Melissa, and my son, Matthew. Also, my gratitude to my parents, Thomas and Janet Lindsey for always encouraging me in everything I do.

BRADY

Fire Service Instructor

Jeffrey Lindsey, Ph.D., CFO, EMT-P

Upper Saddle River, New Jersey 07458

Library of Congress Cataloging-in-Publication Data

Lindsey, Jeffrey.
Fire service instructor / Jeffrey Lindsey.
p. ; cm.
Includes bibliographical references and index.
ISBN 0-13-124557-0
1. Fire prevention–Study and teaching. 2. Fire extinction–Study and teaching. I. Title.
[DNLM: 1. Emergency Medical Services–methods. 2. Fires. 3. Safety. WX 215 L752f 2006]
TH9120.L55 2006
628.9′2′0715—dc22

2004060153

Publisher: Julie Levin Alexander
Publisher's Assistant: Regina Bruno
Senior Acquisitions Editor: Stephen Smith
Assistant Editor: Monica Moosang
Senior Marketing Manager: Katrin Beacom
Channel Marketing Manager: Rachele Strober
Marketing Coordinator: Michael Sirinides
Director of Production and Manufacturing: Bruce Johnson
Managing Editor for Production: Patrick Walsh
Production Liaison: Julie Li
Production Editor: Lisa S. Garboski, bookworks
Manufacturing Manager: Ilene Sanford
Manufacturing Buyer: Pat Brown
Creative Director: Cheryl Asherman
Senior Design Coordinator: Christopher Weigand
Cover Designer: Christopher Weigand
Composition: The GTS Companies/York, PA Campus
Printing and Binding: Courier Westford
Cover Printer: Coral Graphics

Pearson Education Ltd.
Pearson Education Singapore, Pte. Ltd.
Pearson Education, Canada, Ltd.
Pearson Education—Japan
Pearson Education Australia Pty, Limited
Pearson Education North Asia Ltd.
Pearson Educación de Mexico, S.A. de C.V.
Pearson Education Malaysia, Pte. Ltd.

10 9 8 7 6 5 4
ISBN 0-13-124557-0

Contents

Foreword

Training and education are at the core of the emergency services industry. The equation is simple: without training you have injured or dead firefighters. It must be stressed that training alone is not sufficient. The knowledge of why we do what we do—the science behind the skills—is not only essential, it is imperative. Perhaps most important, though, is the assurance that *what* we are being trained is the right thing to do, and that we are being trained the correct way to apply any given application. As fire service instructors you play one of the most important and pivotal roles in the future and ultimate safety of the firefighter. Instructors are mentors, good and bad, and mold new and old thoughts, beliefs, and practices into functions that will live for decades. What you say, how you say it, and most importantly how you actually do what you say will remain in the minds of your students forever.

Keep firefighter safety paramount in your teachings. Nothing should get in the way of safe applications and practices. Instructional subject matter is developed for specific reasons, and alteration of the subject content and context is not an option for the instructor. The mere functions performed every day by the firefighter depict and dictate this need. It has been said by many prominent individuals in various industries that nothing should supersede safety. I offer to you that in the fire service you must take this one giant leap further: *Nothing, absolutely nothing, can supersede safety.* Education and training material is developed and published for clear and specific reasons. As an instructor, you are a representative of the subject matter being taught. Work hard to dispel any mentality suggestive of comments like "Don't worry about what you learned in class, or what the book says, here is how it works in the real world." This mentality is rooted deep in some minds and could not be further from the truth. Such thinking equals unsafe practices that can lead to firefighter injury and death.

I have spent a few of my teenage years and all of my adult life in the emergency services industry. My peers not only taught me basic fire fighting practices, they instilled the importance of safety very early in my career. My first fire chief, John Nelligan, and Deputy Chief Dale Walmer, with the Grantville (PA) Volunteer Fire Department, provided quality sound leadership with a safety culture built into the foundation of everyday fire fighting instruction and practices. Only a teenager at the time, I was taught basics about personal protective equipment (PPE) and the importance of wearing the right protective gear and perhaps most important, wearing that gear the right way. The use of PPE was not the only safety message sent during my early days; an unwritten message stood out as well. That message was that *individual safety* is of the utmost importance. I am not completely sure if these two individuals knew the message they concurrently sent, but rest assured, as fire service instructors, your teaching will be heard, written or unwritten, loud and clear.

Although I grew up in the era where riding the tailboard was not only acceptable, it was considered cool, it did not take away from the safety culture being instilled in a young, impressionable firefighter. Of course, today we all know that riding a tailboard is not only unsafe and unacceptable, it is definitely not cool, but this practice still goes on in some departments. At age 20 I attended my first emergency vehicle driver training class. That was 24 years ago. I still meet people today who have been in the emergency services business as long or longer than I have who have never taken such a course. How can that be?

Each of us is confronted with conscious and subconscious practices virtually every waking moment. On and off the job every function carries an inherent or learned act: getting out of bed, taking a shower, shaving, and picking up a cup of coffee, not to mention the hundreds of learned fire fighting functions. These and many, many more daily habits and functions also occur in the firehouse. Coupled with these daily routines are the extraordinary practices performed by firefighters responding to an incident, on location, and returning from the incident scene.

As you develop your personal styles and techniques with instructor methodology, consciously be aware that not only what you say, but what you do, will be instilled in the firefighter's mind forever. As you study this text and perhaps build lesson plans from its contents remember the effect you will have each time you teach. Until next time... Be safe!

Warm regards!
Richard (Rick) William Patrick

Preface

It has amazed me how we thrust firefighters into a teaching role just because they are good at fighting fire or extricating individuals from vehicles. Just because an individual is good at performing a skill or is extremely knowledgeable on the subject does not make them a teacher. In many cases the opposite may be true.

As I have progressed through my career, I have had a real desire to teach. There were many times that I felt personally responsible for the failure of my students. In many cases this may have been true, but I did not want to settle for being just an okay instructor and wondering if I could have done something better or different.

There are a number of instructor courses that teach you to instruct their program but fail to give you some of the essential components of adult education and techniques for teaching in a truly educational realm. My desire to be a better instructor took me back to the educational setting to earn a master's degree and then a doctoral degree. I by no means advocate that every instructor has to earn a graduate degree; therefore, it is my intent to share with fire service instructors a solid foundation in education theory and experience. It is my sincere hope and desire that those instructors who read this book do not make the same mistakes that I have made. This text includes information that I wish I had known before I began teaching.

Chapter 1 begins by discussing the role of the instructor. An instructor wears many hats and this chapter discusses many of these hats. Additionally, it is essential that instructors are ethical in their dealings in the classroom and with their students. This chapter also discusses discipline, and safety during training.

Chapter 2 gets into the methodologies of instruction. This chapter deals with the art of teaching and the characteristics of learners. Chapter 3 continues with learning theories. Many instructors will want to skim this chapter, but it gives the foundation of how students learn and enables instructors to be better prepared to teach the variety of students in their classrooms.

The classroom environment is critical to create the best situation for the student to learn. Chapter 4 covers classroom layout and design and how to create a conducive learning environment.

Chapter 5 gives the basics in regards to legal aspects. Instructors are not lawyers; however, there are some basic fundamental areas that are essential for the instructor to understand in the legal arena. This chapter gives a good overview, but it is important to always consult your legal department with any concerns.

Live fire training is becoming more difficult every year. Chapter 6 gives instructors the information they need to conduct a live fire burn. In addition, simulation is discussed for creating a live training session.

Testing and evaluation is many times a difficult subject. Chapter 7 gives the instructor a good foundation and an in-depth look at creating a reliable and valid test.

Chapter 8 provides the instructor with an overview of instructional media. Technology continues to evolve and it is important for instructors to stay current in the latest technology.

There are times when instructors may want to develop their own curriculum. Chapter 9 provides a solid foundation in the curriculum development process.

Depending on its size, instructors may have to not only instruct but run their training program. An entire text could be devoted to this subject; however, Chapter 10 includes only the basics of organizing and running a training program.

An instructor needs to continually develop professionally. Chapter 11 gives an overview of some of the opportunities for professional growth along with the resources for these opportunities.

I sincerely hope you gain knowledge and insight from this text. With luck, it will help you avoid some common mistakes. To me, there is nothing more rewarding than teaching: Both education and training are an important part of my life. I wish you the best of luck in your teaching career and challenge each of you to be the best instructor you can be. It is a great way to make a difference in someone's life!

Good luck!
Jeffrey Lindsey

Acknowledgments

Gerald Van Gelder, Plans Examiner and Instructor at Estero Fire Rescue, for enduring the many readings of the chapters and giving me feedback.

Katrin Beacom from Brady for giving me the opportunity to write this text.

Richard Patrick, VFIS, for being a friend and always being there when I need someone to bounce things off, but most of all being my critic when I instruct to make me a better teacher. He is the president and chief executive officer of the Annville (PA) Volunteer Fire Department.

Reviewers

Gene P. Carlson, MIFireE
Red Lion, PA

Larry C. Perez
Associate Professor, MS, CEFI
Las Cruces, NM

Tracy E. Rickman, MPA
Fire Coordinator
Rio Hondo College
Whittier, CA

About the Author

Dr. Jeffrey Lindsey has served in a variety of roles in the fire and EMS arena since 1979. He started his career in the profession as Junior Firefighter in Carlisle, PA. He has held positions of firefighter, paramedic, dispatcher, educator, coordinator, deputy chief, and chief. He has also worked in the insurance industry in education and risk control. Dr. Lindsey serves on various advisory boards, state and national committees, and also writes a monthly column for *JEMS,* a national EMS journal. He is the education chair for the Florida Fire Chief's Association EMS section, a board member for the National Association of EMS Educators, and serves on the state advisory board for education in Florida. He is also a member of the American Society of Training and Development, Florida Society of Fire Service Instructors, and International Society of Fire Service Instructors. He also is involved with emergency vehicle safety and serves on the Federal Interagency Committee on EMS as the International Fire Chief's Association EMS section representative.

He is currently the Operations Chief for Estero Fire Rescue. In 1985 he pioneered the first advanced life support service in Cumberland County, PA. Dr. Lindsey has experience in various environments from the Philadelphia Fire Department, Carlisle Fire Department, Largo Fire Rescue, and currently Estero Fire Rescue.

He holds an associate degree in Para-medicine from Harrisburg Area Community College, a bachelor degree in Fire and Safety from the University of Cincinnati, and a master degree in Instructional Technology from the University of South Florida. He has earned a Ph.D. in Instructional Technology/Adult Education from the University of South Florida. Dr. Lindsey is completing the Executive Fire Officer Program at the National Fire Academy.

He has designed and developed various courses in fire and EMS. Dr. Lindsey is accredited with the Chief Fire Officer Designation. He also is a certified Fire Officer II, Fire Instructor III, and paramedic in the state of Florida. He holds a paramedic certificate for the state of Pennsylvania. He is also a certified instructor in a variety of courses. Dr. Lindsey has taught courses throughout the country and has been a speaker at many national and state conferences including Fire Rescue Med, Fire Rescue East, NAEMSE Symposium, Colorado EMS conference, Emergency Vehicle Technician Symposium, JEMS Technology conference, and others.

The Role of the Instructor

CHAPTER

Terminal Objective

The participant will be able to define the roles of an instructor in a classroom and function as a fire service instructor as defined by NFPA 1041, Fire Service Instructor Professional Qualifications.

Enabling Objectives

- Differentiate between the three levels of instructors as defined by NFPA 1041, Fire Service Instructor Professional Qualifications.
- Define the various roles of an instructor.
- Define the characteristics of an instructor.
- List the responsibilities of an instructor.
- Explain how ethics influence the students and the instruction in a classroom.
- Identify the types of difficult students and how to effectively deal with them.
- Discuss how to issue discipline in the classroom.
- Describe the types of feedback.
- Discuss the instructor's role of safety in the classroom.

JPR NFPA 1041—Instructor I

4-4.4 Adjust presentation, given a lesson plan and changing circumstances in the class environment, so that class continuity and the objectives or learning outcomes are achieved.

4-5.5 Provide evaluation feedback to students, given evaluation data, so that the feedback is timely, specific enough for the student to make efforts to modify behavior, objective, clear, and relevant; include suggestions based on the data.

◆ INTRODUCTION

Teaching others the knowledge and experience gained as a firefighter can be extremely rewarding and satisfying. Since an instructor plays a critical role in delivering information to students and not all fire service personnel are prepared to share their knowledge in an educational setting, the selection of an instructor for the training program is an important one. Many times the individual most knowledgeable about a subject is selected to be the instructor—but breadth of knowledge doesn't necessarily make a great instructor. A person can be the best at what they do, but may be a very poor instructor. Often being a good instructor just means having the proper foundation. This text, which is primarily designed for new instructors, establishes the groundwork for success.

This chapter begins, appropriately, at the very beginning: defining the role of the instructor. It also addresses the qualities of the instructor, ethics and how they apply to a fire service instructor, and the types of feedback and discipline in the classroom an instructor can expect to confront. The chapter concludes with a summary of the instructor's responsibility to promote classroom safety.

◆ FIRE INSTRUCTOR STANDARD

The various levels of a fire service instructor need to be defined. The National Fire Protection Association (NFPA) standard for Fire Service Instructor Professional Qualifications is NFPA 1041, which defines three levels of instructors: Instructor I, Instructor II, and Instructor III. This standard outlines and identifies the competencies required to meet the minimum qualifications for fire service instructors.

NFPA 1041 defines each level of instructor as follows:

> Instructor I—A fire service instructor who has demonstrated the knowledge and ability to deliver instruction effectively from a prepared lesson plan, including instructional aids and evaluation instruments; adapt lesson plans to the unique requirements of the students and authority having jurisdiction; organize the learning environment so that learning is maximized; and meet the record-keeping requirements of authority having jurisdiction.
>
> Instructor II—A fire service instructor who, in addition to meeting Instructor I qualifications, has demonstrated the knowledge and ability to develop individual lesson plans for a specific topic including learning objectives, instructional aids, and evaluation instruments; schedule training sessions based on overall training plan of authority having jurisdiction; and supervise and coordinate the activities of other instructors.
>
> Instructor III—A fire service instructor who, in addition to meeting Instructor II qualifications, has demonstrated the knowledge and ability to develop comprehensive training curriculum and programs for use by single or multiple organizations; conduct organization needs analysis; and develop training goals and implementation strategies.

Essentially an Instructor I is a classroom instructor, who is expected to take a lesson plan and teach a course. An Instructor II, the next level, develops a lesson plan and performs more of a supervisory role to instructors. An Instructor III is considered more of a training program manager. At this level the fire service instructor is running the training program and developing curriculum.

At the beginning of each chapter in the text the Job Performance Requirements (JPR) for each instructor level will be addressed as they are defined in the NFPA standard. These requirements will be identified by the color-coded JPR heading.

CHOOSING A CAREER PATH AS AN INSTRUCTOR

You can travel several avenues as a fire service instructor. The doors are wide open and depend only on the path of travel and level of education of the instructor.

One such avenue is the training officer, whose primary job is to provide the training and oversee the training program for the personnel of an agency. Typically these instructors are mainly concerned with the personnel of his/her department. The training officer may be charged with training recruits, staff, or both.

Another career path can be found in regional, county, or state training centers, which are located in many parts of the country. These centers typically employ full-time personnel to manage the training center and hire contract instructors to teach the programs. These centers will conduct a range of training, from recruit classes to continuing education courses to courses for the general public or business and industry.

A number of private companies conduct training as well. These companies hire instructors on a full-time, part-time, or contractual basis. Training courses come in many different formats and typically teach everything except recruit classes.

Finally, colleges and universities have full-time and adjunct fire service instructors. These institutions may offer anything from recruit classes to continuing education courses and typically offer college-level courses for those seeking an associate, bachelor, master, or doctoral degree. Keep in mind that collegiate instructors usually need at least a master's degree to teach at the associate or bachelor's degree level and a doctorate degree to teach at the master's and doctorate level.

Once the decision has been made on which path to travel, the next step is to choose which role to perform.

A fire service instructor wears many hats. In some ways it's like being a firefighter. Firefighters are called upon to function in different capacities, from helping an elderly person back into bed, to testing hydrants, to fighting a structure fire. An instructor will also function in different roles, depending on the position and the agency. (See Figure 1.1.)

The various roles of a fire service instructor include:

Administrator
Counselor/Adviser
Disciplinarian/Coach
Evaluator
Facilitator
Mentor
Presenter
Representative/Ambassador
Role Model
Supervisor

Let's examine these various roles.

FIGURE 1.1 ◆ A fire service instructor teaches in many settings and uses different tools to get the message across to students.

ADMINISTRATOR

In this function the instructor may be responsible for overall course operations and logistics. An effective administrator must be organized, detail oriented, and have effective communication skills. Often an instructor must orchestrate a number of diverse personalities, and in the role of the administrator it becomes even more evident.

Part of an administrator's duties include scheduling and planning a course along with arranging the appropriate facility for the course. This latter means determining if the room is large enough for the activities and if the room needs to be configured in a certain way to provide the best delivery.

In any organization communication is always a critical element. For an instructor in an administrative role, it's essential to initiate communication with the students prior to class using flyers or other physical means. This flyer should give the course title, a brief description of the course, plus its date, time, and location. It's always a good idea to require registration to ensure that the class isn't overcrowded and there are enough materials. Confirmation letters should be sent to the students, which should give the location, time, and day of the class. This makes it easier for the student to confirm their registration and remind them of the course's location.

Record-keeping is another essential administrative element that an instructor must perform. Primary among these records is the course roster that students must sign for documentation of class attendance. Other documents, too, typically need to be completed by the student and filed or returned to the director of the training programs. These may include course evaluations and tests. Record-keeping is discussed later in the text.

COUNSELOR/ADVISER

Students often look to their instructor for guidance. An instructor may serve as a counselor or adviser, and an open and trusting relationship depends on showing that as the instructor, he/she is trustworthy, empathetic, and an active listener. Being discreet and keeping a student's confidence are vital to building and earning trust. This, in turn, will ensure that the student will be open and honest when discussing any type of issue with the instructor. (See Figure 1.2.)

A counselor offers guidance and assistance to the students within the instructor's role. When problems or issues affect the classroom or a student's performance, it is entirely appropriate for the instructor to discuss the situation with the student and work with him/her. A student's ability to function in a classroom setting can be affected by issues at work, school, or home. In many instances, the student feels comfortable confiding only with the instructor. The instructor may find that just listening is enough. If that isn't the case, the instructor needs to know when his/her limit has been reached. At this point refer the student to a professional therapist or other form of counseling appropriate to the situation. An instructor can't be all things to all people. Instructors need to know their limitations.

FIGURE 1.2 ◆ There are times when an instructor needs to counsel students. This should be done as tactfully and diplomatically as possible. *(Photo courtesy of James Clarke, Estero Fire Rescue)*

DISCIPLINARIAN/COACH

Discipline is too often deemed a negative term. *Coaching* might be a better term. As a coach, an instructor establishes—and requires compliance to—standards of behavior. To be effective in this role, the instructor needs to both consistently enforce these behavioral standards and convey the consequences for noncompliance.

Disciplining effectively isn't simple—but remember: It is easier to become more lenient than to get tougher. A good coach is very clear about what is expected and equally clear about the consequences of noncompliance, especially since the field of fire service requires rising to a higher level of standard than many other occupations. New recruits must be ready for the responsibilities of the job; if the recruits are not ready, an instructor needs to identify these problem areas during training and effectively guide the students in the right direction. In some instances, this may mean directing the recruit toward a new career path. This is the coaching end of being a disciplinarian.

On the other hand, the instructor may have difficult students in the classroom—students who, for whatever reason, are disruptive and uncooperative. This situation calls for the more tyrannical aspect of being a disciplinarian. Instructors must demonstrate that they are ultimately in charge. The best way to control difficult students is to be respectful yet assertive and direct. How the instructor handles a situation can affect the attitude of all the participants and potentially the success of the course. (The different characteristics of difficult students and how to handle them will be discussed later in this chapter.)

EVALUATOR

Every class should have a set of terminal and enabling objectives, in many instances derived from a set of standards. It is the evaluator's role to compare the performance of the student against the standards. A variety of formal and informal methods can help evaluate a student's progress, including:

Written and oral tests and quizzes
Essay questions
Practical exams
Project assignments
Observational reports
Presentation checklists
Peer review
Question and answer sessions

The evaluator's responsibility is to conduct a fair evaluation of each student. Chapter 7 will go into more depth regarding evaluation techniques and strategies.

FACILITATOR

It is the instructor's role to aid or assist the student in the learning process. To accomplish this, the instructor uses facilitation skills so that students feel free to comment and ask questions.

Facilitative teaching emphasizes student involvement in the learning process—that is, students are not just passive recipients of the presentation. (See Figure 1.3.) Consequently, students influence the delivery of material more than in traditional, lecture-only formats. The upside to student-paced delivery is a likely increase in comprehension. The downside is that scheduling can be affected. An instructor needs to

FIGURE 1.3 ◆ A facilitator does not dominate the classroom, but creates an atmosphere to get the students actively involved in the learning process.

monitor the discussion and the time involved when facilitating a class. In many instances course content will need modifying or schedules will need changing to compensate for the increased amount of discussion.

Adult students tend to prefer a facilitated class over a traditional lecture class. An adult student has a wealth of experience and knowledge that he/she enjoys sharing, and a facilitated class permits this type of environment.

MENTOR

A mentor assists personal and professional growth. A good mentor also recognizes each student's strengths while being aware of and accepting his/her weaknesses and vulnerabilities. Typically, good mentors are high achievers who are confident in their own positions. Conversely, they don't feel threatened by students who excel and achieve in the classroom. Mentors motivate students and encourage professional growth. As a fire service instructor, be proud to be in the profession and look for ways to promote and expose the students to new opportunities and challenges in the fire service. (See Figure 1.4.)

As a mentor, be careful not to favor certain students. Though it's natural for a relationship to be established in a mentoring format, creating favorites can be very detrimental.

PRESENTER

Presentation—which is the most visible part of what an instructor does—plays an extremely vital role in a classroom. The more an instructor can make a presentation come alive, the easier it is to capture and keep the students' attention. Some instructors

Figure 1.4 ◆ A mentor is there to assist the learner. In many instances the mentor is in the background encouraging and directing the learner, allowing the learner to gain experience firsthand. *(Photo courtesy of James Clarke, Estero Fire Rescue)*

relate presentation to being on stage and feel a responsibility to maintain a high level of energy in the classroom. Whether this kind of performing exhausts or energizes the instructor, it's important to know that instructors give it their best throughout the entire course.

REPRESENTATIVE/AMBASSADOR

Serving in the role of instructor makes the instructor the ambassador or representative for the organization. As a representative, be professional and have high qualifications. Whether the instructor is representing his/her organization or another organization, it is the instructor's responsibility to communicate the agency's message appropriately, following prescribed guidelines.

The instructor needs to be well versed not only in the specifics being taught for the organization, but state and national standards as well. Whatever the agency has accepted as the standard for the course that is to be taught is the material that needs to be delivered. Deviating from this standard opens many liability issues for the agency and the instructor. In many instances the instructor will be given a lesson plan or set of guidelines, and it is his/her responsibility to fill in the content.

As an instructor, enhancing the material being taught may be appropriate. The instructor should not delete any of the content or objectives that have been established as the core of the course material. It is the instructor's responsibility to meet a course's established objectives.

Trainer's Toolbox

Invariably, one of the first things a student asks when first walking into a classroom is, "When are we going to be done?" The instructor's reply should be, "When the course objectives are met." The design of the course should be such that the time allocated meets the objectives set for the class. If need be, amend the instruction to fulfill the lacking component—whether the time allotted is not enough to get all the material covered or too much time to cover the material.

ROLE MODEL

An instructor is also a role model. This means being an achiever, having a positive attitude, and being admired by others. Hard work, dedication, and a genuinely caring manner are other desirable traits of a fire service instructor. As the saying goes, imitation is the sincerest form of flattery—and since an instructor's behavior is typically imitated by students, the instructor should maintain high standards, both professionally and personally.

SUPERVISOR

A supervisor directs and inspects performance. Being responsible for students, the instructor is essentially the supervisor. Additionally, a fire service instructor will need assistant instructors from time to time, especially when practical evolutions are conducted. Initially the instructor may not be responsible for other instructors, but eventually this, too, will fall under the purview of the instructor's responsibility; and it's imperative that the lead instructor recruits assistant instructors or other instructors who are qualified and capable of conducting the course. (See Figure 1.5.)

GETTING HELP

There will be times when a new instructor needs help. Don't be afraid to ask for assistance when needed—even seasoned instructors seek help from time to time. In fact, the instructor should expect that the employer and supervisor will assist in the growth and development as an instructor. Seek out a mentor who will guide and help. The mentor does not need to be a fire service instructor. Choose someone who has a good rapport with students, has an excellent delivery technique, and is well respected by other instructors and students.

INSTRUCTOR QUALITIES

Take a moment, and think back over your educational experience. No doubt there were certain instructors or teachers that had a great influence. What do you remember about these individuals that made them so good at their job?

Particular traits separate a great instructor from a merely adequate one. First and foremost, an instructor needs to genuinely love the subject. Instructors should not teach a subject they can't relate to! Instructors will not be doing the students justice by presenting content that they don't agree with, don't understand, or just don't like. Let someone else cover this topic.

Another characteristic of a good instructor is preparation. Always arrive early to ensure that the room is set up and capable of handling the course logistics. This

Figure 1.5 ◆ Instructors do not need to know everything there is to know, but need to surround themselves with assistants and adjuncts to compliment their skills and abilities. *(Photo courtesy of James Clarke, Estero Fire Rescue)*

includes having the appropriate AV equipment in working condition. (Media instruction is discussed in Chapter 8.) An instructor also needs to make sure that any needed handouts have been duplicated well in advance of the class. Make sure that all easel charts are set up and hands-on instructional materials are present. The last thing the instructor wants to be is disorganized—or perceived by the students as such.

A good instructor:

- Is well prepared. There is no acceptable excuse for an instructor not being well prepared.
- Is enthusiastic about the subject.
- Cares about students.
- Uses appropriate body language, gestures, and eye contact.
- Uses positive feedback.
- Uses appropriate humor.
- Dresses appropriately.
- Utilizes a variety of inflections and pauses for effect—*not* "um's" and "uh's."
- Is punctual—both starting and ending class.
- Practices innovative teaching methods.
- Talks with students, not to or above them. Intersperses appropriate real-life illustrations.
- Is professional in all ways.

- Is relaxed and comfortable within a classroom setting.
- Answers all questions honestly and with regard to whom they represent.
- When stumped by a question, readily admits to not knowing the answer and is willing to find out.
- Stays calm and does not become defensive.

What additional characteristics would you add to this list?

Remember: The instructor represents the agency when presenting the course. The students and the agency are relying on the instructor's ability to deliver the message that has been designed in the curriculum. Behaviors that should never be employed as an instructor include:

- Yelling at students.
- Berating or belittling students.
- Ignoring concerns for safety issues.
- Attempting to motivate students with threats.
- Encouraging or permitting off-color humor or sexist comments.
- Being unprepared with course material.
- Showing nervousness: twirling hair, jingling change, clicking pen caps, etc.
- Showing personal bias and nonsupport of instructional material.
- Communicating information ineffectively.
- Chewing gum or tobacco.
- Consuming alcohol or illegal drugs before or during the class time.

Here are a few do's:

- Make eye contact with all participants.
- Present the course material without reading from the manual.
- Relate "war stories" on a limited basis with relevance to material.
- Arrive early for the session.
- Use terms at the participants' level.
- Move around the classroom to keep students engaged, using appropriate gestures.
- Use nongender-specific words such as "firefighter" or "emergency responder" instead of "fireman."

Most of all it is important to *be yourself!* Don't try to be someone else, no matter how much you admire the person's style. Consider observing a good instructor in action, and adapt what he/she does well to fit *your* style. Remember: You are who you are. As hard as you may try, you cannot be someone else—nor do you want to be someone else. An instructor should find an instructional niche—and then improve upon it.

One of the most beneficial ways to improve is to teach using the buddy system. This means finding someone that can be bluntly honest; conversely, that person will need to be bluntly honest with you.

TRAINING TALES

Early in my teaching career, I found a person to be bluntly honest with me. My partner and I would team-teach a course, and after the course was over, we gave each other constructive criticism. In many cases we would be brutal, not holding anything back. It is amazing to learn the things you do without ever realizing that you do them. As a footnote: We started doing this fifteen years ago, and we are still friends today, doing this same exercise not only in teaching but in other areas as well. Incidentally, though it's not as effective as having a buddy pointing out your weak areas, videotaping yourself is another means of identifying your shortcomings.

RESPONSIBILITIES AND ACTIONS

An instructor needs to promote a positive, dynamic class and provide training to enhance performance. As an instructor these are the responsibilities and actions:

- Set a constructive, comfortable classroom environment.
- Facilitate interaction, and avoid lecture-only presentations.
- Understand the importance of following the course materials.
- Complete and submit all required paperwork in a timely fashion.
- Motivate your students.
- Be a role model during as well as outside of class.
- Correct errors in technique using positive reinforcement.
- Reinforce students' strengths and efforts to succeed.
- Guarantee a physically and emotionally safe learning environment.
- Keep students' attention on the goals. (See Figure 1.6.)

FIGURE 1.6 ◆ A fire service instructor does not always have the luxury of being in a clean, comfortable classroom. There are times that group discussion is an important element of the learning process, even in the environment pictured here. This group was discussing the activities of rescuing individuals in concrete rubble. *(Photo courtesy of James Clarke, Estero Fire Rescue)*

ETHICAL ISSUES

An ethical person behaves in accordance with the established moral principles that govern the conduct of a group. An instructor must behave ethically both to establish trust in the student/teacher relationship and inspire confidence. An instructor can lose integrity far quicker than it took to gain it. Build moral credibility by acting with:

- Fairness
- Integrity
- Honor

Definitions

Exactly what *are* ethics, morals, and values? A number of definitions are acceptable for each of these terms. Though basically ethics is generally thought of as the study of right action, and morals is the system through which that action is applied, the definitions go deeper than that.

Ethics. Ethics can be defined as the critical examination and evaluation of what is good, evil, right, and wrong in human conduct. Ethics is also the study of goodness, right action, and moral responsibility, asking what choices and ends we ought to pursue and what moral principles should govern our pursuits and choices. Ethics can also be a specific set of principles, values, and guidelines for a particular group or organization.

Morals. Morals are principles and values that actually guide, for better or worse, an individual's conduct. Morality is the informal system by which rational beings govern their behavior to lessen harm and do good. The concept of morality—its rules, idea, and virtues—enjoys amazing agreement across time and cultures.

Values. Values are the guiding principles of behavior and conduct; where emphasis is placed; and what is rewarded in an organization and society.

With these definitions in mind, let's take a deeper look at ethical issues as they relate to being an instructor and as they relate to subject matter in the class.

ETHICS

Ethics is an interesting topic. Entire courses are devoted to ethics, and there is no better time to discuss this topic than when fire service recruits are first entering the profession. If instructing a fire recruit class, incorporate the topic into the class. By doing so, it makes a statement to the students about the importance of ethics in general and its integral importance in the fire profession. Do some research to locate some of the numerous books and articles dealing with this subject.

Activity

Think about the many times that an ethical question has surfaced during your career, and write down examples. Before long you will have a list of ethical questions that you can use in the classroom.

Determining right from wrong and what is ethical in today's world is debated more than ever. A facilitated environment allows the students to discuss these issues.

As an instructor, create an atmosphere that is a teaching environment, not a preaching atmosphere. Adult students in particular dislike being preached to and will typically shut someone out if they feel this is occurring.

ETHICAL THEORIES

Having an in-depth knowledge of ethical theories isn't necessary to help shape classroom discussions, but there is a need for a general understanding of some of the major theories.

The first is *divine law,* which is based primarily in Judeo-Christian and Islamic religions. According to divine law, if something adheres to God's will and word, it is considered good. Furthermore, if someone is obeying God's will, he/she will exhibit the right behavior. This theory provides moral certainty and guidance, which are its strengths. In contrast, the weaknesses of this theory are that, depending on ones beliefs, divine law emphasizes moral certainty, self-righteousness, and intolerance.

The next theory is *virtue ethics,* which is based in the ancient Greek philosophy of Plato and Aristotle. According to virtue ethics, "good" is seeking happiness and living the good life, and right behavior is exhibited by acting virtuously, which is necessary for happiness. The strength of this theory is that virtue is its own reward and leads to self-actualization. The weaknesses are that consequences, common good, and principles are ignored.

The third theory is *egoism,* which is based in classical and contemporary philosophy. According to this theory, what is considered good is what I think is best for me, and right behavior is defined as promoting what is good for me alone. The strengths of this theory are that it leads to moral certainty and moral autonomy. Its weaknesses are self-centeredness, moral certainty, selfishness, and unrealistic thinking.

The fourth theory is *ethical relativism,* which is also based in classical and contemporary philosophy. What is considered good according to ethical relativism is whatever the individual/group/culture decides is good, and right behavior is acting in accord with the group's values and principles. The strengths of this theory are tolerance of others, flexible thinking, and practicality. The theory's weakness is that it rules out criticism of obvious evil and everything is considered relative.

The next theory to discuss is *utilitarianism,* which is based in British/American philosophy: Bentham and Kant. "Good" is happiness/pleasure, diminishing misery, and pain, while right behavior is the promotion of the greatest good for the greatest number. This theory's strengths are that it is practical and considers the consequences of actions. Its primary weakness is the rationale that a good end may justify a bad means. This often vague theory justifies dehumanizing mistreatment of a minority group of people as the means to an end if they do not agree with the majority.

The last theory to highlight is *duty ethics,* which is based on theories by Kant. Goodwill that is good-hearted and extended to others is considered good; right behavior is doing your moral duty and acting as a model for others to follow. The strengths of this theory, which is consistent and certain, are the promotion of highly principled behavior and showing respect for self and others. Unfortunately it ignores circumstances and principles and offers no way to choose among competing principles.

Guidelines for Leading a Discussion on Ethics

After introducing ethical concepts and theories to the class, present a case study or ethical issue. This is a great way to start the group discussing ethical problems. Consider dividing the students into groups, giving each group the same scenario to see how the different groups react to finding responsible answers or solutions to the problems posed. Figure 1.7 is a guide to some classroom activities that you can conduct involving ethical issues.

Activities and Classroom Exercises

Resources of ethical issues

1. Internet sites
2. Print-based materials
3. Current events
4. Actual fire calls and scenarios
5. Colleges and universities
6. Ethical think tanks and centers

Group discussion/debate

1. Present a case or scenario of an ethical issue.
2. Divide the class into groups, and assign one of the theories just presented to each group for discussion.
3. After 15–20 minutes have each group defend a particular ethical theory as it relates to a case or scenario presented.

Case scenarios

1. Provide several short case scenarios, and ask students to do the following:
   a. List the options for each scenario.
   b. State what decision you think is the right one.
   c. Explain why they think the decision is the right one.
   d. Explain what theory they are aligned with in making their decision.
2. Allow students 5–10 minutes to think about each scenario before comparing their responses with the class.

Role-playing
Role-playing will help students gain perspective by experiencing another point of view.

1. Choose a case study, and have volunteers role-play various sides of the issue to present their side of the story.
2. If the instructor has prior knowledge of a student's opinions on certain issues, consider having the student role-play an opinion that is diametrically opposed to their personal feeling and beliefs.

Debates

1. Present a case, and allow students some time to prepare their viewpoint on the issue.
2. Conduct the session as a debate, allowing students to challenge and defend their opinions.

FIGURE 1.7 ◆ Suggestions for activities and exercises that can be used in the classroom. *(National Guidelines for Educating EMS Instructors (August 2002) National Highway Safety Transportation Administration)*

The key to answering ethical issues is knowing when and where to ask the right questions. Use the following list of questions as a guide.

1. What are the facts of this particular case?
   Do I have everything I need to know, or am I acting on rumor?
   Am I letting bias or emotions distort the facts?
   Is this primarily a legal or policy issue, not an ethical one?
2. Who is involved?
   Who is responsible for causing this issue or problem?
   Who is responsible for deciding what to do?
   Who will be harmed or helped by the actions taken?
3. Why have I chosen the ethical action I have?
   What values and principles am I basing my decision on?

As with any discussion that may get emotional, ground rules need to be established at the beginning of the discussion. It may be advisable to write these somewhere in the classroom so that everyone can be reminded of them. This will set the tone for the discussion. (See Figure 1.8.)

ETHICAL ISSUES IN TEACHING

A number of ethical issues surround teaching. Primary among them is plagiarism. Be sure to give credit where credit is due. Falsifying documentation is another key issue. Accurate record-keeping is an important element of instruction. Emphasize that cheating will not be tolerated. Taking risks when working for real in the fire service is an interesting topic to discuss. If a firefighter makes a daring rescue and the outcome is positive, he/she is typically rewarded for a heroic effort; however, if someone got hurt or possibly killed, he/she then becomes the subject of discipline. Where is the line drawn between a heroic action and being foolishly risky?

The environment in the firehouse setting is more contentious than it has ever been. Unethical or inappropriate language or behavior with other firefighters, families, staff, or the public is unacceptable. This includes the act of hazing. Unacceptable classroom behavior, such as violence, threats, or harassment, is another ethical issue the instructor needs to recognize. Some individuals have a difficult time transitioning to the environment behavior that is expected of them.

Dealing with Ethical Issues in Teaching

The first thing an instructor can do to foster a positive learning environment is minimize behavior problems. An instructor is a role model. This includes modeling ethical

Ground Rules for an Ethical Discussion

1. Everyone who wants to speak may do so.
2. Students will respect each other's differences of opinion—in other words, they can agree to disagree.
3. Students will be polite to each other.
4. Students must back up their opinions with the facts as they see them, not just spout opinions.

FIGURE 1.8 ◆ Ground rules for discussions on ethics.

behavior both in as well as outside the classroom. A firefighter is held to a higher standard. An instructor is held to an even higher standard. This encompasses everything from appropriate dress and language to demonstrating concern and respect for others. As a fire service instructor, a commitment to academic excellence and lifelong learning is to be made.

Additionally, the agency should publish classroom rules, policies, and expectations so that every student and instructor will be on the same page. Establishing the ground rules for all to know ensures a better chance of maintaining a well-behaved and -disciplined classroom. Discipline will be discussed in more detail later in this chapter.

DEALING WITH PROBLEM STUDENTS

On occasion the instructor may have a difficult student in the classroom. These individuals can easily distract from the intent of the class. One way to handle the situation would be to have a private conference. Being gentle but firm, tell the disruptive student to squelch his/her behavior, warning that further disruption will not be tolerated, and dismissal from the class may be necessary. If the instructor is an officer conducting the class, it may be easier to command the attention of a group and gain the control of the class. If there is an officer in the class, use his/her assistance in keeping class control. Occasionally, despite using the preceding tactics, controlling the class becomes impossible. Under these extreme circumstances dismiss the class—and the message that behavior matters becomes clear when the class finds out that there will be no credit for the hours attended. This should be used only as a last resort, and a superior needs to be notified immediately of the actions along with written documentation of the incident according to the Standard Operating Procedures (SOPs).

The following are some of the more common difficult students an instructor may encounter.

The Hesitator. The hesitant person is shy, reluctant, and silent most of the time. Strategies for dealing with this student include using small group activities, calling on them from time to time to answer nonthreatening questions (ones you know they can answer), and offering encouraging statements that let them know that their contributions are worthwhile and appreciated.

The Monopolizer. The monopolizing student tends to be opinionated and likes to dominate class discussions. This type of student can dampen the enthusiasm of the other students, who may need clear openings and encouragement to participate. Some statements to use with this type of student are: "I'd like to get another opinion on this issue" or "I appreciate your input, but everyone needs an opportunity to participate."

The Voice of Experience. Closely associated with the monopolizing student, the know-it-all has a tremendous need to be heard, as well. This person likes to display their knowledge to everyone by using big words, lots of statistics, even occasional name-dropping. Always be polite, but maintain control of the discussion by moving to the next topic. If this person is knowledgeable, give them a task or even a leadership role. Another tactic is to administer a test that is difficult but is at the level the student claims to be. When it's made clear that there are things he/she doesn't know, the student will more than likely be ready to learn.

The Nonlistener. At times a student's attention may wander. One strategy to refocus a nonlistener is to ask something in this manner: "Could you take what Stephanie just said and explain it another way?" Or ask, "How does your viewpoint compare with what has been expressed?"

If multiple students seem to be "tuning out," it could be a cue to the instructor that a break is needed, the instruction is not sinking in, or there's an environmental distraction that needs to be handled.

The Idea Zapper. This person is expert at putting down the ideas shared by others in the class or finding creative ways to inhibit suggestions or cast doubt on solutions. This can seriously undermine small group interactions as well as classroom discussion, so be sure to watch for the idea zapper as groups break out to work on their assignments.

During discussions, rescue an attacked idea before the whole group dismisses it by making concrete statements that confirm potential usefulness. Then, ask the idea zapper to come up with his/her idea.

The Master of Negativity. Fault-finding gripers exist just about everywhere. Stop a complainer in his/her tracks by asking questions that force the student into a problem-solving mode, such as, "What steps do you feel are necessary to correct this situation?" It can also help to say "I understand" occasionally, depending on the reason for the negative attitude.

The Rigid Thinker. This type of person will take a position on an issue and does not want to budge—that can make it difficult for the group to make progress. Try to get this type of person to admit that there is another side to every issue. One strategy is to ask the student to clearly state the rationale behind his/her opposing viewpoint.

The Antagonist. Some students can be hostile, aggressive, and unfriendly. Fortunately, adult learners are typically strongly motivated to be involved in instruction, so bad attitudes are relatively rare. The exception is that adults might be more likely to contest results that prevent them from achieving their goals, such as a too-low test score that prevents certification. If the problem is administrative or grade-related, refer the student to the appropriate grievance procedure.

If, however, you have an inexplicably angry student in your classroom, avoid getting wrapped into a debate. Remain calm and respond in a mild, objective manner. Sometimes activities can redirect energy toward accomplishing a specific task. There is nothing like success to turn a negative attitude around.

The Class Clown. This type of student hinders group progress with an abundance of inappropriate humor. Strategies for dealing with clowns include complimenting them when they make a worthwhile contribution and never rewarding the inappropriate humor with laughter. During a serious dialogue, remember to ask this individual to contribute.

The Slow Learner. Occasionally a student will have trouble keeping up with the group or may not be able to grasp a certain part of the material. The instructor can approach this student in a variety of ways to overcome this obstacle. Allow input from the other students, who may be able to explain the topic in question in a manner that the slow learner can easily understand. Sometimes hearing something said a different way might be enough for the student to comprehend the material.

Sometimes a student who is a slow learner may not be getting the message, despite everyone's best efforts. Don't allow the student to slow the entire class to a point that the rest of the group becomes lost. Instead, consider giving the class a break and talking with the student one on one. The instructor can then advise the student of his/her interest in the student's learning, but simply don't continue to hold back the rest of the class. Perhaps suggest that the student could attend the rest of the class and either discuss the areas of concern after class or attend another class in the future. Do not embarrass or belittle the student. The student may become ambivalent toward the instructor, the class, or the agency the instructor represents.

LEAST

Somewhere during my teaching career, a student gave me this mnemonic to remember how to deal with difficult students: LEAST.

Leave it alone.
Eye contact.
Action steps.
Stop the class.
Terminate. Let's look at these in order:

Leave it alone. If it's the first occurrence, it may well be an isolated incident, and if ignored, it may just fix itself. The participant may just be looking for attention and when not given it, the student will cease the actions on his/her own.

Eye contact. If the problem occurs again, make direct eye contact with the problem student. By looking disapprovingly at the student, he/she may get the message and cease the disruption. Be careful not to be too condescending or overreacting when using this step.

Action steps. If the problem behavior persists, give your presentation with the problem student in close proximity. By being in the student's space, he/she may understand his/her disruption and cease the behavior.

Stop the class. It may be a good time for a break. Stop the class, take a break, and speak with the problem student about his/her disruptiveness.

Terminate. Expelling the problem student from the class. If matters reach this drastic level, notify the superior immediately.

As an instructor, treat the adult learner the way you would want to be treated in the classroom setting. Whenever participating in a course, think of how the instructor should conduct the class. Evaluate the positive aspects of the class, and the instructor shouldn't be afraid to adopt them in his/her instruction delivery if it's appropriate to his/her teaching style and personality. Remember: The adult learner expects a lot from the instructor.

◆ DISCIPLINE

Earlier in this chapter the role of the instructor as a disciplinarian or coach was discussed. Let's examine this aspect of instruction more closely.

CLASSROOM DISCIPLINE

Unacceptable classroom behaviors disrupt the learning process and may even pose physical danger to the instructor or other students. As a fire service instructor in a training institution, there may be a legal liability to provide an appropriate classroom environment. However, depending on the infraction, disruptive students may still have legal rights, and it's important for instructors to know how to appropriately handle classroom and student problems.

Unacceptable Classroom Behaviors

Unacceptable behaviors can be grouped into two categories: those that are considered illegal (criminal or tort) and those that are uncomfortable (disruptive or undesirable but not clearly criminal or tort). Illegal behaviors include: violence, threats of violence, sexual harassment, hazing, discrimination, or destruction of property. (These will be discussed more thoroughly in Chapter 4.) Uncomfortable behaviors include: foul language, being loud, an angry tone, sleeping, or nonparticipation.

Cost of Uncontrolled Classrooms

An uncontrolled classroom can be very damaging to instructors, on both emotional and professional levels. Behavior management is the leading cause of career stress for teachers and the most common reason teachers leave the profession. Classroom management also affects how others perceive the competence of an instructor. These include the students, colleagues, and administrators (fire chief, operations manager, dean). An uncontrolled classroom limits the time to teach and learn. (See Figure 1.9.) It also leads to an unsafe and negative learning environment.

The Root of Behavior Problems

The following are likely causes of behavior problems:

1. Poor parenting
2. Lack of societal values
3. Anonymity in large institutions
4. Boredom

FIGURE 1.9 ◆ The classroom will have many disruptions; students talking with one another occurs very frequently. It is important to keep order in the classroom for the benefit of all students.

5. Substance abuse
6. Economic situations
7. Lack of recognition
8. Family stress
9. Poor coping skills
10. Poor communication skills
11. Lack of social skills
12. Weak institutional policies and penalties

Correlations Between Behavior and Cause

A number of correlations can be derived between behaviors and the cause, such as the following:

- If you are annoyed, the student is probably seeking attention.
- If you feel threatened, the student is probably seeking power.
- If you feel hurt, the student is probably seeking revenge.
- If you feel powerless, the student is probably seeking adequacy.

Examples of Correlations

Let's take a look at some of the examples of these correlations.

Seeking Attention. In this instance a student calls out all the time, asking irrelevant questions. They also give excessive examples.

Seeking Power. This student may be exhibiting tantrumlike or quarrelsome behavior. They will frequently lie and may refuse to follow directions.

Seeking Revenge. This correlation is typical with students who are cruel to others. This student will actually try to get punished—and in the same respect will dare the instructor to punish him/her. It is not unlike these individuals to cause vandalism or perform pranks.

Seeking Adequacy. Students who feel inadequate passively refuse to participate in any classroom activity. They sit silently and never speak or interact. If an instructor calls on them, they don't answer. They ask not to be included in any activity or discussion. (See Figure 1.10.)

CREATING POSITIVE BEHAVIORAL CHANGES

Creating positive behavioral changes starts by preventing behavioral issues. Anticipate any behavioral issues by having written rules telling students what is expected. These rules should be included in the student manual or student syllabus. List all the consequences, from mild penalties to removal from the classroom or program. This information should be reviewed with the students in the beginning of the course or program and revisited periodically if problems arise. The instructor should also include information on grievances in the manual; students need to be aware of and understand their rights as well as their responsibilities. In some instances, it's a good idea to require students to sign a document attesting that they have received the information and know what was expected. One copy of this signed form should be given to the student for his/her records, and another copy should be entered in the student's file. (See Figure 1.11.)

If rules are established, do the homework ahead of time and be sure that the rules don't contradict those of the facility, the program, or the state. Submitting the plan to administration for approval will guarantee administrative support should the need present itself to enact the final phases of discipline.

Figure 1.10 ◆ Some students will want to sit by themselves. It is important to try to get them involved and pulled into the class environment, too. Using other students usually works best to accomplish this.

Figure 1.11 ◆ Documentation and record-keeping are an important function of the instructor. There will be many times that you will need to complete various paperwork as part of the instruction responsibilities.

STEPS TO TAKE IN THE CLASSROOM

Good behavior begins with strict (though fair) rules and regulations. It is easier to lighten up than tighten up. Do not allow the students to be intimidating. A common mistake in this situation is disciplining the student as a result of that intimidation. Be aware of this situation and avoid it.

It is also important to watch for opportunities to reward students' good behavior. If operating a fire recruit academy, utilize class leaders—elected by their peers—for peer policing of unacceptable behavior.

Furthermore, it can't be emphasized too much that a fire service instructor is a role model and is being watched constantly by the students. The instructor always needs to demonstrate courteous and respectful behavior.

The instructor also needs to be organized and prepared for each class to minimize distractions and lost time. Be respectful of the students and realize that in most instances, the students are paying money to be in the class. The students are also taking time away from other things to be in the class. Start on time, and end on time.

Also remember that it's good to see the humor in situations. There are times to laugh. This is especially true when *the instructor* does something wrong. Students need to know that the instructor can make mistakes and not necessarily be criticized for it. Instructors need to be responsible for their actions: however, everyone makes mistakes.

Finally, never plead with the students to behave. That only shows a weakness on the instructor's part. If students aren't behaving, and the class is out of control, it may be time to at least take a break. The instructor should gather his/her thoughts, regroup, and reconvene. If things still do not go well, it may be time to dismiss the class and talk to the supervisor about how to regain control of the class. It might mean having another instructor assigned to the class. Always try to identify the cause of the behavior before acting to correct it. Gather facts before jumping to conclusions about the incident. Jumping to conclusions will not make the instructor good in the classroom or put the instructor in good stead with his/her superiors.

IMPOSING DISCIPLINE

As an instructor consistently enforce rules by moving through the consequences in progression. If necessary seek assistance from other members of the education team, including the program administrator or coordinator, the fire chief, and other faculty such as primary and secondary instructors. Having a mentor to consult with would be invaluable.

When dispensing punishment, utilize the principles of progressive discipline. These actions in the disciplinary process include a reminder, a verbal reprimand, a counseling session, removal of privileges, a written warning, a suspension, and, as a last resort if warranted, removing the student from the class or program.

Situations that involve illegal activity or threaten the safety of others necessitate immediate removal from the classroom setting. In these instances, inform the supervisor, and get his/her involvement.

Be sure to take the necessary steps so that the discipline will stick if challenged. It is very frustrating when you need to reinstate a problem student because a step in the process was neglected. The instructor needs to respect a student's right to due process. The student is allowed the opportunity to have legal representation and to present an alternate perspective to the situation. This should all be spelled out in the grievance procedure.

When the instructor metes out discipline, do it in private. It is recommended, however, that another individual be in the room to witness the disciplinary action. (See Figure 1.12.) In cases where the instructor is disciplining a student of the opposite sex, someone of the same sex as the student should be a witness. A witness should be

FIGURE 1.12 ◆ When counseling students, it is a good idea to have window blinds open and a witness in the office.

another instructor either in the fire service program or another discipline, though avoid having the supervisor as a witness in case the student decides to grieve the process. In most instances when a student grieves a discipline that has been issued, he/she will grieve it to the next level, which is typically the instructor's supervisor.

Document all infractions to establish a pattern. The following need to be included:

- Time and date
- Any witnesses
- Description of the incident(s)
- Unacceptable behavior
- Corrective action taken

Provide documentation to the student, and inform him/her who will receive copies of this information. Full disclosure is not only fair, it may be enough to stop the behavioral problem. Attempt to discover the cause of the behavioral problem to address the underlying issue, not just the surface symptoms. It is important to protect the privacy of the student involved, and to address the root of the problem; but when appropriate, utilize other services such as:

- Employee Assistance Program
- Counselor
- Physician
- Tutor
- Student health services

FEEDBACK

Feedback can be an essential tool to an instructor. The three types of feedback are positive feedback, constructive feedback, and corrective feedback.

Positive Feedback

Positive feedback is a great morale booster and reinforces desirable behavior. Positive feedback removes doubt, builds self-esteem, and results in a sense of accomplishment. Positive feedback is given when a student does something right—and an instructor can almost always find something someone does right. Some instructors find it hard to get in the habit of giving positive feedback and instead want to criticize or critique what someone does. It may take a concerted effort on the instructor's part to give positive feedback, but the more he/she does it, the more it will become rote.

Constructive Feedback

Constructive feedback helps change undesirable behavior. When giving constructive feedback, avoid using statements or questions that put the student on the defensive and instead express feelings. For example, instead of asking a student when he/she is going to quit being late for class, the instructor should tell the student how annoyed he/she feels every time the student is late for class.

When giving constructive feedback:

- Describe the specific behavior for which you are giving feedback.
- Don't use labels such as "immature" or "unprofessional."
- Don't exaggerate; that heightens emotions.
- Don't be judgmental; that produces defensiveness.
- Use "I feel" statements.
- State the consequences calmly and clearly.

Corrective Feedback

Corrective feedback is used to improve student performance incrementally. This technique involves analyzing performance, identifying correct and incorrect components, and communicating specific information that the student can use to make subsequent performance improvements. This should be a positive learning experience and part of the ongoing, informal evaluation process. Corrective feedback lets students know where they stand, thereby reducing frustration and tension in the classroom. It also prevents students from assuming everything is just fine—that "No news is good news." This prevents problems later, during formal evaluations.

When giving corrective feedback:

- Be descriptive.
- Be specific.
- Be private; students hate to be embarrassed in front of their peers.
- Be positive; find something good in every performance.
- Be concise; give information in manageable chunks.
- Be timely—immediately post performance if the student is ready to listen.

CREATING A SAFE TRAINING ENVIRONMENT

Trainer Tales

Marine fire fighting session injures instructor

In 1998 a firefighter incurred burns to his hand in the course of a marine fire fighting training session. The firefighter was using gasoline, which led to a vapor explosion when he removed a glove and lit the training fire. Although marine fire fighting is not addressed in NFPA 1403, the individual survived with just minor injuries primarily because he was wearing full turnout gear and SCBA that is required by NFPA 1403. Lately, though, many fire training incidents have had more serious and even fatal consequences.

It is the instructor's responsibility—and the instructor sets the precedent—to create a safe environment for the instructor, the students, and the other instructors. Training should not result in injuries, and certainly a fatality should not result from a training evolution. Granted, things happen, but in the training setting mechanisms should be in place to prevent or mitigate any potential misfortunes from occurring.

STUDENT INJURY

Fire fighting is a dangerous occupation. Training can be just as dangerous. Each year a number of individuals are injured and in some instances killed as a result of training exercises. (See Figure 1.13.) Some will say that this is part of the job and the risk

FIGURE 1.13 ◆ Training scenarios are no different than real life. There are many instances when students have been injured or killed during training exercises. The same precautions need to be exercised during a training evolution as at a real incident.

FIGURE 1.14 ◆ Safety needs to be the number one priority during a training evolution. This picture illustrates a live training burn where arrows were painted in fluorescent orange to mark the exit paths out of the building.

firefighters take. But one injury or one fatality is one too many. Injuries can occur as a result of student or instructor error. It is the instructor's responsibility to ensure a safe environment. Instructors should not subject the students to any training evolution that is known to be unsafe. Supervision needs to be in place at all times. A training exercise is a time for students to learn and receive guidance to prevent any mishaps from occurring. This includes using inadequate, malfunctioning, or faulty equipment. It is imperative that the instructor check all the equipment prior to the start of class. If the equipment is not working or not adequate, don't do the evolution.

A safe environment includes both the physical and the emotional. Remember: A safe environment begins with the attitude of the instructor. What the instructor says and what he/she does is what the student remembers. It is always a goal to create an environment as real as the one a firefighter will encounter. But it is not a goal to create an environment so real that potential for serious injury or death is a possibility. Training should be designed to teach the student how to handle a certain situation within a learning environment. Whether conducting a driver-training class or staging a live burn, safety is an instructor's number one priority. (See Figure 1.14.)

◆ SUMMARY

An instructor wears many hats. From an administrator to an evaluator, the roles encompass a variety of responsibilities. An instructor should possess many qualities, including being ethical and moral. Those same values need to be instilled on the students. An instructor also needs to be a disciplinarian, and must be prepared to issue fair and consistent discipline. Another primary task of an instructor is to provide feedback. Finally, an instructor must create and maintain a safe environment.

The profession of teaching is a rewarding and fulfilling career. Whether full-time or part-time, there is no better compensation than watching others succeed using what the instructor has taught them.

Review Questions

1. List and describe the roles of an instructor.
2. Describe the different types of feedback. Discuss how you would use each type of feedback effectively.
3. Discuss how you would deal with a difficult student in your classroom.
4. Describe the process of delivering discipline in the classroom.
5. List the qualities of a good instructor.

Bibliography

Aiken, T. D. 2002. *Legal and ethical issues in health occupations*. Philadelphia: W. B. Saunders Company.

Coughlin, S., Soskolne, C., and Goodmath, K. 1997. *Case studies in public health ethics*. Washington, D.C.: American Public Health Association.

Edge, R., and Groves, J. 1999. *Ethics of health care* (2d ed.). New York: Delmar Publishers.

Madden, T. 2000. *A compendium of ideas and resources for using ethics across the curriculum*. Howard, MD: Howard Community College.

National Highway Safety Transportation Administration. August 2002. *National guidelines for educating EMS instructors*.

Silberman, M. 1998. *Active training*. San Francisco, CA: Jossey-Bass/Pfeiffer.

Thiel, A., Stern, J., Kimball, J., and Hankin, N. 2003. *Trends and hazards in firefighter training*. (No. USFA-TR-100). Emmitsburg, MD, Federal Emergency Management Agency.

Methodologies of Instruction

CHAPTER

Terminal Objective

The participant will be able to describe the various methods of instruction and adapt it to a classroom setting.

Enabling Objectives

- Differentiate between the art and science of teaching.
- Define andragogy.
- Describe the characteristics of an adult learner.
- Describe how to motivate the adult learner.
- Describe the skills of making an effective presentation.
- Describe the four major teaching strategies.
- Describe how to conduct a facilitated class.
- Describe how to conduct a Socratic seminar.

JPR NFPA 1041—Instructor I

4-2.2 Assemble course materials given a specific topic so that the lesson plan, all materials, resources, and equipment needed to deliver the lesson are obtained.

4-3.2 Review instructional materials, given the materials for a specific topic, target audience and learning environment, so that elements of the lesson plan, learning environment, and resources that need adaptation are identified.

4-3.3 Adapt a prepared lesson plan, given course materials and an assignment, so that the needs of the student and the objectives of the lesson plan are achieved.

4-4.4 Adjust presentation, given a lesson plan and changing circumstances in the class environment, so that class continuity and the objectives or learning outcomes are achieved.

4-5.5 Provide evaluation feedback to students, given evaluation data, so that the feedback is timely, specific enough for the student to make efforts to modify behavior, objective, clear, and relevant; include suggestions based on the data.

JPR NFPA 1041—Instructor II

5-3.2 Create a lesson plan, given a topic, audience characteristics, and a standard lesson plan format, so that the job performance requirements for the topic are achieved, and the plan includes learning objectives, a lesson outline, course materials, instructional aids, and an evaluation plan.

5-3.3 Modify an existing lesson plan, given a topic, audience characteristics, and a lesson plan, so that the job performance requirements for the topic are achieved, and the plan includes learning objectives, a lesson outline, course materials, instructional aids, and an evaluation plan.

5-4.2 Conduct a class using a lesson plan that the instructor has prepared and that involves the utilization of multiple teaching methods and techniques, given a topic and a target audience, so that the lesson objectives are achieved.

JPR NFPA 1041—Instructor III

6-3.4 Modify an existing curriculum, given the curriculum, audience characteristics, learning objectives, instructional resources, and agency requirements, so that the curriculum meets the requirements of the agency, and the learning objectives are achieved.

◆ INTRODUCTION

If instructors are teaching a public education class at a school or in the community, students will typically be adults in the 18- to 70-year-old range. Because adults learn differently than children, it's important to understand how to best teach an adult so that the instructor presentation can be adapted accordingly. This chapter describes andragogy, or adult methodology of learning. It also discusses the motivation of adult learners, different types of presentation styles and instructional strategies, and how to facilitate a class. (See Figure 2.1.)

THE ART AND SCIENCE OF TEACHING

Teaching is said to be both an art and a science. The art of teaching involves presentation style, presence in the classroom, and rapport with the students. These areas are not based on any scientific theory, but rather on the art of instruction. This chapter discusses the art in teaching. The science of teaching includes learning styles, learning theories, and adult education. (See Figure 2.2.) Chapter 3 will discuss the science of teaching.

FIGURE 2.1 ◆ There are different methodologies for teaching individuals. The traditional classroom may not be the best method. Interactive teaching is becoming more of the norm in today's classroom.

ADULT METHODOLOGY OF LEARNING

The Father of Adult Education, Malcolm Knowles, adopted a theory that is referred to as *andragogy,* or the adult methodology of learning. Knowles's theory emphasizes that—unlike children—adults are self-directed and are expected to take responsibility for their decisions. An instructor needs to accommodate this fundamental as part of the training program in the course and classroom.

The following assumptions are made about the design of learning in andragogy:

- Adults need to know why they must learn something.
- Adults need to learn experientially.
- Adults approach learning as problem solving.
- Adults learn best when the topic is of immediate value.

To put the term *andragogy* to practical use, incorporate strategies like case studies, role-playing, simulations, and self-evaluation into the delivery of training programs for the classes and the classrooms. (See Figure 2.3.) Andragogy evolves around the role of the instructor being a facilitator as much as a teacher. Facilitation will be discussed later in this chapter.

FIGURE 2.2 ◆ All instructors have their own comfort zone and method of instructing.

FIGURE 2.3 ◆ Adults enjoy hands-on activities and typically learn best in this format. Teaching fire behavior at a live fire burn is usually more effective than teaching it in the classroom.

The principles Knowles uses in applying his theory have been adapted as follows for the fire service instructor:

1. Explain why specific things are being taught (for example, incident command, hose evolutions, hydraulics, etc.).
2. Instruction should be task-oriented instead of memorization—learning activities should be in the context of common tasks to be performed.
3. Instruction should take into account the range of students' backgrounds; learning materials and activities should allow for different levels and types of experience with using computers.
4. Instruction should allow adult students to discover things for themselves, providing guidance and help when mistakes are made.

Characteristics of Adult Learners

Adult students have many traits inherent to their learning ability. The following is a general description of some of these abilities.

Adults are generally autonomous and self-directed learners. Unlike grade school or high school days, when the focus is mostly on the teacher, adult students function best if the environment is centered on them, not on the instructor. Therefore, choose the presentation style carefully. For example, a lecture tends to be instructor-centered while small group activities, are student-centered.

An adult does not like to be told what to do. As children we were always told what to do and how to do it. Adult learners need to be free to make their own decisions. This does not mean a free-for-all in the classroom; the class is guided rather than directed. This is where the art of facilitation comes into play. In a facilitated classroom the students have the feeling they are in control, but the instructor facilitates the discussion, keeping things in perspective. A good facilitator will listen to what the students' needs are and direct the discussion toward the feedback being received from the students.

Trainer's Toolbox

When you begin a class it is very helpful to go around the room, have students introduce themselves, and ask each of them what they want out of the class. Then ask why they are in the class. You may be surprised at some of the answers. In some instances, it's because the student was told to take the class. Being told to attend a class is not always a great motivator, but at least you'll be aware of why the student is there. Once the participants have told what they want out of the class, tailor the class toward the expectations. The instructor needs to meet the objectives and follow the lesson plan, of course, but tailor the discussion and the course to the needs of the student. You'll have more interactive, happier students who'll get more out of the class.

If you do this exercise, write the responses on large sheets of paper, and post them around the room. At the end of the class look at each of the responses and quickly determine if indeed the topics were addressed. This is almost an informal process of creating a list of course objectives that's great for closure and demonstrating to the students that the instructor really cares about what they want.

Adults have gained a foundation of life experiences. Regardless of their age, each of the students has had past experiences—both good and bad. And these experiences can have a major impact on adult learners. For example, the instructor may resemble a teacher in his/her past with whom he/she had a terrible experience, creating, unbeknownst to the instructor, a negative barrier that the instructor must overcome. Indeed, the student may not even be consciously aware of his/her negativity or why he/she does not like the instructor.

Activity

Take a few moments to reflect back on the various classes you have attended. Now try to recall the other students who attended the class. What characteristics stood out most about them? What about the teacher? How was the instructor perceived by the students? Did he/she contribute to the students' attitude? How would you have done things differently if you were teaching the class? Share your responses with the students or with other instructors and see what *their* responses would have been.

An instructor needs to connect the students' knowledge/experience base to the learning in the classroom. Have students draw on their experiences in the classroom. Adults also are very critical about textbooks—and the fire service probably even more so. As an instructor it's great to relate real-world experiences to the topic being taught. When used properly the experiences can be very effective in showing the relationship of why students are learning what they are learning. Be careful, though; these experiences can be negatively construed as war stories if used in excess.

Trainer's Tip

I cannot count how many times I have heard a firefighter tell a "rookie," "Forget what you learned in the classroom; it is different in the real world." That is true to a certain degree, but theory or book learning sets the foundation for what the student will face on the job. (See Figure 2.4.)

Trainer's Tip

I had an instructor who once said that a bad example can actually be a good example because it shows a person what not to do.

Adults are relevancy oriented and practical. Adult learners want to know why they need to learn this or why they need to take a class on that. If the instructor can show relevancy, the student will be motivated to learn. Additionally, the instructor needs to show how the learning will benefit students when they leave the classroom and go back to the real world. If students do not think there will be any relevancy or practical use to what the instructor is teaching, they probably will not place as much importance on being part of the class.

Adults need to be shown respect. As the instructor, the first thing to realize is that you don't know everything there is to know about any given subject. There will be areas or particular subjects where the fire service instructor is the expert and knows more than just about anybody in the classroom—and for that matter may know more than most in the profession; however, there are still things to learn. Therefore, it's important to recognize what instructors don't know and the wealth of experience the

Figure 2.4 ◆ It is important to conduct practical exercises illustrating the points taught in the classroom. This photo depicts a final practical exercise after the completion of hazardous materials (hazmat) training. *(Photo courtesy of James Clarke, Estero Fire Rescue)*

students in the class bring with them. Yes, the instructor will stand in the front of the classroom, but the instructor needs to treat the students as peers, not as subordinates.

Trainer's Tip

In just about every class I teach, I learn something. Sometimes I think I learn more than the students. Encourage the students to share their opinions and experiences in the classroom. The next time you teach this same topic, you will have gained valuable information from the students in the previous classes that will make the next class even better.

Adult students usually want to utilize the knowledge and skills soon after they have learned them. Adults typically enjoy hands-on activities, so allow the adult students to have a hands-on session and apply the information they have just learned. An Emergency Vehicle Driver Training program is an example of a program that accomplishes this principal. Driving on the competency course is usually the highlight of this program. It lets the students put their classroom knowledge to actual use. Most students will typically say that they would rather be doing the hands-on practical evolution than sitting in a classroom.

Adult learners are interested in learning new concepts and principles. Adults enjoy situations that require problem solving rather than simply learning facts. When teaching or designing courses, be sure to include student interaction. Class participation lets students gain more from the course. Granted, not everyone will agree with everything the instructor says; however, the goal is to have participants begin to think about how they do things. Does what they do make sense or can they do it a better way? Adults typically learn more if they are active participants rather than passive listeners.

Motivation is increased when the content is relevant to the immediate interests and concerns of the students. Each geographic area has its own unique characteristics. Sometimes the instructor needs to adapt the course material to meet regional needs. By no means, though, should the instructor change the content or the intent of the course material to make it fit a personal bias. For example, if the instructor is teaching pump operations, and the students in the class are from a rural fire service where there are not any hydrants, the time should be spent on tanker shuttles and drafting versus hydrant operations. This is not to say that hydrants should be completely left out of the program—just spend more time on topics relevant to the needs of the student.

Immediate feedback is essential to the adult learner. Students need to be kept informed of their progress. Positive reinforcement is a great motivator when giving feedback to a student. If the instructor is not accustomed to providing positive feedback, the instructor needs to think about when he/she is in the role as a participant, and think about what type of feedback helps to improve. If the instructor continually sends negative reinforcement messages to the students, eventually the students will be turned off by them and will think they aren't capable of learning what the instructor expects of them. If an instructor accentuates the positive by continuous positive reinforcement, students will be more positive about the class and will probably learn more.

MOTIVATING THE ADULT LEARNER

One of the keys to being an effective instructor is motivating students. Motivation is the key to getting students involved and becoming active participants in the education process. Instructors need to be motivated so they can motivate others. The instructor may want to determine the motivation level of each of the students. As suggested earlier in the Trainer's Toolbox, this can be done at the start of the class by having students introduce themselves and identify their primary motivation for taking the course. This will enable the instructor also to identify the reason for positive or negative behavior the student may demonstrate during the course of the class and help to plan activities that build intrinsic motivation. Essentially, as instructors present the material in class, they need to keep in mind what students identified during their introductions. The instructor can use a variety of icebreakers to introduce the students. The following sidebar offers one common icebreaker that's also a means to learn about each of the students.

Trainer's Toolbox

At the beginning of the class, have the students pair up and interview each other. They should get the person's name, where they are from, why they are in the class, and any other information you think is appropriate. Give them 5 to 10 minutes to gather the information, and then have each introduce the other. This may not be appropriate for every class, but it is a great ice breaker for many courses.

Intrinsic motivation comes from within. This includes the desire to help others. Many emergency workers will tell the instructor that they got into the fire service to help others and perform a community service. Instructors will tend to hear this as motivation in the introductory fire service courses they teach. Typically the students have not been involved in the service for any length of time. Unfortunately it seems quite prevalent for some individuals to lose this desire after they have been in the fire service for a while.

Personal growth and development is another intrinsic motivation that students will cite for being in a class. These students want to improve on their skills to advance their knowledge and career. The instructor will also find a number of these students who have a strong drive to succeed. The great thing about students who have a high level of intrinsic motivation is that they help motivate other students in the class.

Extrinsic motivation comes from outside rather than inside. Extrinsic motivations include such things as money, time off from work, and job requirements. These motivators can be good or bad, but they are motivation for the student to be in the class.

The following are other areas to consider for motivation:

- Social relationships: to make new friends or meet a need for association or friendship
- External expectations: to fulfill the expectations of someone of authority
- Social welfare: to offer community service and to serve humankind
- Personal enhancement: to achieve higher status at work, provide professional advancement, or stay abreast of competitors
- Escape/stimulation: to relieve boredom, provide a break from the routine at home or work, or provide contrast to the exacting details of life
- Cognitive interest: to learn for the sake of learning, seek knowledge for its own sake, or satisfy a curious mind

There are also a number of barriers to motivation. They include:

- Lack of time
- Lack of money
- Lack of confidence
- Scheduling problems
- "Red tape," bureaucracy, or politics
- Problems with child care
- Problems with transportation

Be aware that barriers do exist, but also understand that the instructor can do something about these barriers. In Chapter 1 the various roles of the instructor were discussed: mentor, guide, and advocate. The best way to motivate adult learners is to enhance their positive motivations for enrolling in the course and decrease the negative barriers that arise.

Create an environment that aids motivation. It is important to set an appropriate level of stress in the classroom: not too high and not too low. Fire service classes can promote higher stress because the student will eventually be responsible for rescuing others and protecting properties. Instructors tend to create a higher stress level to emulate the level of stress that the student will face in real-world scenarios. To a certain degree this is an acceptable practice; however, the classroom is also the place for students to learn a procedure or operation, and adding stress—at least in the beginning—isn't necessarily a positive motivator.

PRESENTATION SKILLS

The instructor is the key to presenting the material effectively. Indeed, in many instances the success of a program depends on the instructor presenting the material in the most effective manner. It is important to identify the most common instructional styles and be able to use them effectively.

Making the Presentation

The first step is to introduce the subject matter early in your presentation. The introduction should include:

- Your credentials/experience/knowledge
- Description of the course content
- The importance of the material

The instructor also may need to provide motivation to the students. To do this the instructor may need to show the relevance of the material to their work or personal lives.

The next step is to review the course outline. Briefly describe the content, which lets students know what to expect from the course. In a course that spans multiple days, weeks, or months, this knowledge will help the student judge if they want to continue or drop the course. Students also need to know what will be required of them to successfully complete the course. There should be no surprises during the course. Any rules and regulations should be set forth in the beginning. The instructor can also establish any other ground rules for the class. For example, instructors can discuss when breaks will be given or if they prefer questions during or after the presentation.

Appropriate Use of Barriers When Teaching

Many instructors feel more comfortable sitting at a desk or standing behind a podium. (See Figure 2.5.) This is OK, but it can also impart a negative impression to the students. Think about where the instructor is in relation to the student. Is he/she hiding behind objects? Can the students in the back row see the instructor? Does the instructor want to look casual or formal? How approachable or friendly does the instructor appear by where he/she is standing? Generally the instructor wants to stand about eight feet away from the first row of seats. Try to move around the room. Disruptive students are less of a problem if the instructor moves closer to where the student is sitting.

Speaking in Public

The first rule for speaking in public is always use appropriate language. Always use the student's name when addressing them, and never use any other reference when addressing students. The instructor should never use obscenities in the classroom, even when among peers. It is unprofessional, offensive, and may alienate a student. The instructor needs to speak in a clear and distinct voice. If an amplification device is available, use it so that the students can easily hear and speak a little slower than in a normal conversation so that students can understand and comprehend the material.

Consider telling a joke to open the class. It will get people to relax and typically lightens things up. Beware, though: Humor can fall flat if used inappropriately. The fire profession is prone to dark humor as a means of dealing with overwhelming tragedy and relieving stress. Be careful in the classroom. The students may not appreciate this type of humor or other humor that may offend them or others. This is also true when it's the students doing the joke telling.

Eye Contact

It is important to maintain eye contact with the class. Make sure not to hold anyone's gaze for too long, though, since this can make individuals uncomfortable. Watch for

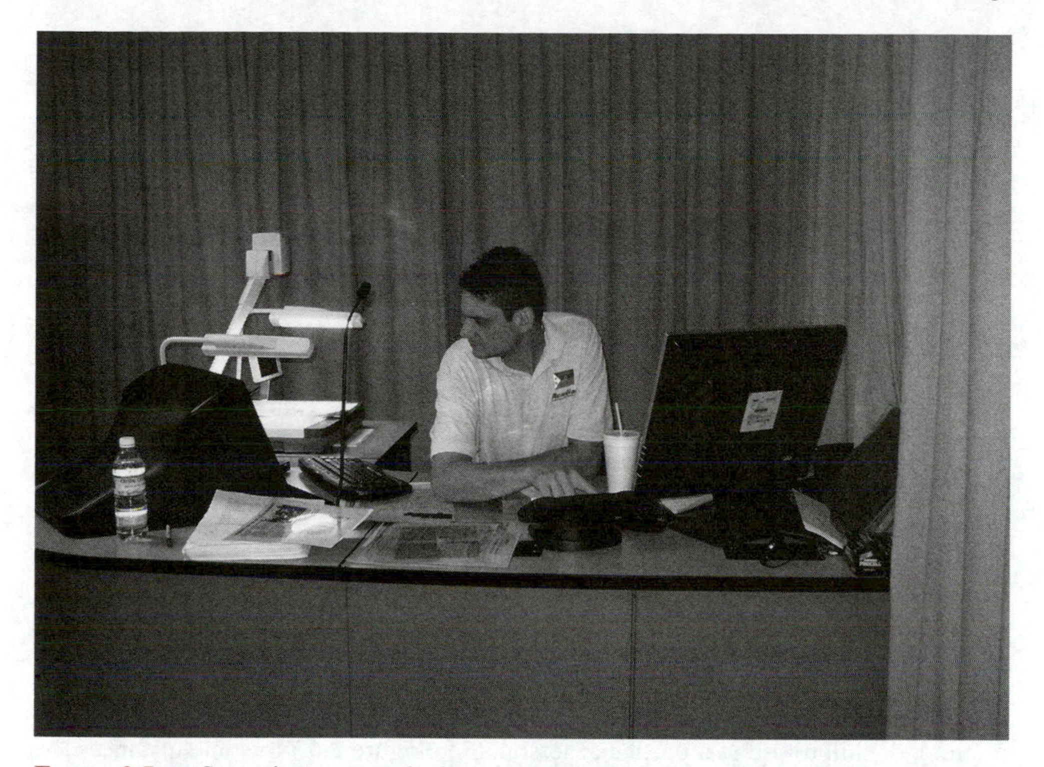

FIGURE 2.5 ◆ Some instructors feel more comfortable lecturing behind a podium. In some instances it is hard to avoid. This classroom is part of an interactive classroom that broadcasts to five other locations throughout Louisiana. Through satellite television, each of the five locations receives the same instruction that is being conducted at the main location. The instructor is confined to the podium in order to operate the technology associated with the interactive classroom setting.

personal blind spots (places you tend to look all the time) because instructors can unintentionally ignore students who are not in their normal vision area.

Body Language

The spoken message conveys less than 35 percent of communication between two people. The unspoken message, body language, conveys more than 65 percent. Therefore, it's important to improve people-reading skills—and equally important to develop understanding of nonverbal communication.

Feelings expressed in body language are influenced by the situation, the setting, the culture, and the individuals. Nonverbal communication is an inexact science, but research indicates that certain feelings and emotions are consistently expressed by specific body movements. Rule of thumb: Words express ideas; body language expresses feelings and attitudes.

A few examples of nonverbal signals are:

Eyes: contracted pupils = concentration
dilated pupils = tuning out
narrowed eyes = intensity or anger
widened eyes = astonishment
eye contact = interest

Mouth: tight lips = anger
open mouth = wonder
crooked mouth = doubt
smile = happiness
frown = sadness
Head: tilted = caring
Chin to chest = aggressiveness
Hands/arms: open = warmth
fist = aggressiveness, anger
hand to chin = concentration
pointed finger = threatening
finger to mouth = perplexed
hands on hips = authority
Shoulders and trunk: slouch = tired, weary
shoulders back = confidence
leaning forward = interest
leaning away = disinterest

PRESENTATION STRATEGIES

The mission as an instructor is to present the course material. The most important component of presenting the material is planning which instructional strategy, method, or skill to use in a particular lesson. (See Figure 2.6.) In some instances the lesson plan will suggest this to the instructor. This section provides various instructional strategies for the fire service instructor. Select those that match the teaching

FIGURE 2.6 ◆ In some settings, lecture is still the most appropriate means for the classroom.

style and the students' learning style. Here are some items to think about when planning the presentation:

- Keep in mind the kinds of students who will be in the class and their desires to learn.
- Adapt instructional strategy to meet the needs of the student.
- Vary the instructional strategies throughout the course or class.
- Keep learning styles in mind. Include written and oral directions and information.
- Try to incorporate practices that involve students in decisions regarding their learning.
- Include enrichment and mediation in different formats.

As mentioned, students will typically be diverse in learning styles; therefore, select instructional methods that make it possible for all participants to learn.

Two commonly used teaching practices are explicit teaching and implicit teaching. In explicit teaching, the teacher serves as the provider of knowledge. In implicit teaching, the teacher facilitates student learning and creates situations in which students can discover new knowledge and ideas. Four major teaching modes can be identified within these two instructional methods. These are:

- Expository
- Inquiry
- Demonstration
- Activity

The following tables are adapted from the *Adaptive Dimension* website by Saskatchewan Education. They outline specific teaching techniques used in the various modes, with suggested adaptations for each technique.

Expository Mode. This is the teaching strategy that is most often used by educators.

Teaching Techniques	Adaptations
Lecture	• Provide lecture outlines. • Provide copy of lecture notes. • Use transparencies or PowerPoint slides to provide visual presentation
Telling	• Keep lecture short. Be specific about information given. • Be sure you have students' attention. • For students with short attention spans, give information in small segments.
Sound Filmstrip	• Provide visuals when possible. • Give earphones to students easily distracted by ambient sounds.
Explanation	• Keep explanations simple and direct. • Give them in simple declarative sentences. • Provide outline of explanation.
Audio Recordings	• Present recordings with visuals. • Give earphones to students easily distracted by ambient sounds.
Videos	• Orient students to video before showing. • Be sure length is appropriate. • Place students with auditory problems close to sound.

	• Review main points of film. • Provide brief outline of main points.
Discussion	• Ask questions you know students can answer. • Keep discussion short. • As points are made, list them on the board, easel, or transparency. • Divide class into groups for brief discussions. • Keep students on topic. • Involve everyone on appropriate levels. • Use organizer to group ideas and show conclusion drawn.

Inquiry Mode. This involves asking questions and seeking information and allows more teacher-student interaction. For students who require adaptations, added teacher involvement is critical in this type of learning situation. Asking questions is a natural part of the teacher's instructional method; in adapting for students' needs, be very aware of the level of questions being asked, since they should reflect the specific learning level of the student.

Teaching Techniques	Adaptations
Asking Questions	• Use appropriate wait time. • Ask questions on appropriate level of Bloom's taxonomy scale; vary questions to meet different taxonomy levels of students. • Call the student's name before directing a question to him or her. • Do not embarrass students by asking questions they cannot answer.
Stating Hypotheses	• Have students choose from two or three hypotheses instead of having to formulate their own. • Provide model for writing hypothesis.
Coming to Conclusions	• Present alternative conclusions. • List information needed for conclusions.
Interpreting	• Assign peer tutor to help. • Present alternative interpretations.
Classifying	• Use concrete instead of abstract concepts. • Provide a visual display with models.
Self-Directed Study	• Give specific directions about what to do. • Make directions short, simple, and few. • Collect and place resources for study in one area.
Testing Hypotheses	• Assign peer tutor.
Observing	• Give explicit directions about how and what to observe. • Provide sequential checklist of what will happen so that student sees steps. • Have student check off each step observed.
Synthesizing	• Assign peer tutor to help. • Provide model of whole or completed assignment.

Demonstration Mode. This involves showing, doing, and telling. Modeling is one of the most effective methods of teaching for students who require adaptations. Models may include maps, charts, globes, or verbal examples.

Teaching Techniques	Adaptations
Experiments	• Provide sequential directions. • Have student check off each completed step. If teacher demonstrates, let student assist. • Be sure student fully understands purpose, procedures, and expected outcome of experiment. • Set up incidental learning experiences. • Display materials. • Model the activity. • Provide an outline and a handout/checklist. • Make a list of lab procedures, and assign a lab procedure. • Tape instructions and videotape demonstrations.
Exhibits	• Assign projects according to student's instructional level. • Have student select project topic from a short list. • Provide directions and list of materials needed. • Be sure project does not require skills the student lacks. • Have students display their exhibits.
Simulated Events	• Do not embarrass students by requiring them to do something they cannot do. • Make sure student understands directions, terms used, and expected outcome.
Games	• Design games in which acquisition of skills, not winning, is the priority. • Make directions simple. • Highlight important directions with color codes. • With peer tutor, let student prepare own game. • Design games that emphasize skills needed by student.
Modeling	• Model only one step at a time. • Use task analysis on steps. • Use visual models when possible. • Exaggerate the presentation to make the concept being modeled clear. • Use several short time spans rather than one long demonstration. • Model in hierarchical sequence. • Use video modeling for student to replay. • Perform in same manner as the first presentation. • Provide a lecture outline that the student can take notes on.

Field Trips	• Prepare students by explaining destination, purpose, expected behavior, and schedule. • Provide a checklist of expectations.

Activity Mode. Students learn by doing. Teachers provide students with actual experiences of concepts.

Teaching Techniques	**Adaptations**
Role-Playing	• Be sure student understands role. (See Figure 2.7.) • Short lines or no lines at all may be best. • Respect privacy of student who does not want a role. • Let such a student assist another role-player.
Constructing	• Select project for students or have them select from a short list. • Try to use projects that include special education objectives. • Provide sequential checklist.
Preparing Exhibits	• Assign peer tutor to help. • Use alterations suggested for "constructing."

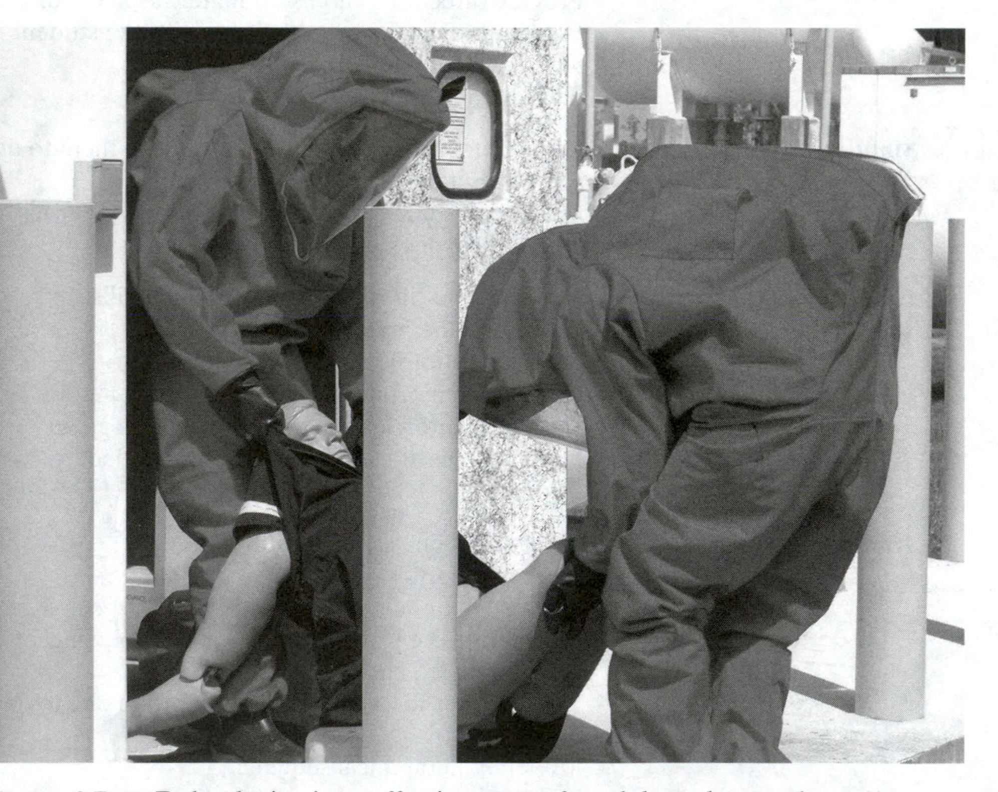

FIGURE 2.7 ◆ Role-playing is an effective means for adult students to learn. An instructor needs to decide whether to use manikins or real patients. If there is any concern for safety, a manikin should be used. *(Photo courtesy of James Clarke, Estero Fire Rescue)*

Dramatizing	• Respect privacy of those who do not want parts. • Let such students help others prepare sets and so on.
Processing	• Clearly state steps. • Make steps sequential and short. • List steps on board.
Group Work	• Assign peer tutor. • Select activity at which student can succeed. • Use variety of grouping procedures.
Game/Contest	• Be sure game matches lesson objective. • Check game to see if required decision-making skills match student's ability level. • List rules clearly on board. • Keep the pace appropriate. • Assign a buddy. • Provide feedback for game skill as well as for social skills used.

Activity

Develop a plan using multiple instructional strategies for a given topic. Include the activity mode, teaching technique, and its adaptations. Present it to the rest of the group.

FACILITATION

Facilitation has become an integral part of today's classroom. The term *facilitate* means to make easier. Facilitation makes the learning experience more productive and enjoyable.

Facilitation is a method of interacting with students that enhances learning. A variety of techniques are used in this setting, including coaching, mentoring, and positive reinforcement. Other terms describe the facilitated learning environment. They include experiential learning, constructivist learning, and invitational learning. The first step in being effective in the art and science of facilitating a class is to know and understand the audience.

Keys to Facilitation

The main key to a facilitated classroom environment is to create action. This means minimizing lecturing and engaging students in learning through activities. The step that is most instrumental in creating a facilitated learning environment is designing the classroom layout to set the tone. The various classroom designs will be examined in Chapter 4. Facilitation creates an expectation in students that they will participate in learning. Often students have been conditioned to be passive learners, which makes it difficult for them to transition into a facilitated learning environment. As the instructor/facilitator, be patient and provide guidance and positive reinforcement to these students. As students succeed, they will change their classroom expectations and adapt to the facilitated environment. Students really need the assurance that the classroom is a safe and nonthreatening environment. Some will still continue to be passive learners

FIGURE 2.8 ◆ A typical classroom setting

despite the instructor's best efforts. Don't be discouraged; eventually those students may want to participate, or other students may influence them to participate.

Lecturing is as old as teaching. This time-honored technique, which focuses on the instructor, continues to be recognized as a method to disseminate a lot of information quickly, with a great emphasis on the instructor having control. Many instructors enjoy lecturing—though often this is because the instructor doesn't want to relinquish the control in the classroom. (See Figure 2.8.)

Unfortunately, lecturing typically does not lead to active learning. It will take some practice, but move beyond simple lectures. Do this by building interest in the material being presented—and the easiest way to build interest is to get the group involved in the topic by centering on something relevant to them.

Trainer's Tip

When I travel around the country presenting courses, I like to read as much as possible about the area and the agency where I will be instructing. Going to an outside agency to teach a course is often the result of an event that occurred, and they realize they need more training in this particular area. I like to arrive early to the class, which a good instructor does regardless, to talk to the students as they arrive. This allows me the opportunity to identify any underlying issues, either as an organization as a whole or on an individual basis. The key to a facilitated learning environment is to get the students involved. As the instructor you also need to reinforce what has been presented in the class.

Let's take a look at some ways to add more interest to the lecture environment. Most people enjoy stories. Instructors can start out by telling the class about an incident that they were personally involved with. The incident may not even be true; however, true incidents add more credibility. The story should relate to the forthcoming lesson to add relevance to the story.

Another way to start off the class is to present a visual aid, such as a picture of a fire or someone doing some type of task. Ask the students to discuss what they see. Have them tell you how they would do things differently.

Or start the class by asking a number of test questions. This gets students' minds in gear and thinking about the class.

Trainer's Tip

When I teach the Training Officer Academy course that I have developed, I like to start out with a 20-question test that has nothing to do with being a training officer. This makes the students begin to think outside the lines of where they have been up until this point. The test is loaded with trivia that even the best *Jeopardy* player would have difficulty answering.

After the class is in full motion, maximize understanding and retention by saying less and letting the students interact more. Accomplish this by giving the *Reader's Digest* version of the subject: Give students the headlines. The student should have read the material prior to coming to class so there won't be a need to recite all the details of the subject.

Alter the presentation to give the highlights in lecture form to the whole class, and then divide the students into small groups to do reinforcement activities with several instructors. Adding some visual appeal will greatly enhance your presentation. Be careful, though not to go overboard with visual effects. This can have as much negative effect as not having any. Providing handouts can also reinforce pertinent points.

Involving participants in the presentation can help to reinforce the important points. It also lets the students put their learned concepts to work. Reinforcing activities should be spaced throughout the lecture. Instead of teaching all the points, assign various parts of the coursework to the students, and allow them to research and present the material. The students will take ownership in the presentation and gain more from the class than by you lecturing on the material. When assigning topics, give the students adequate time to prepare. Instructors need to know the material well enough so that if a student gets off on a tangent or presents wrong information, they can tactfully correct the mistakes and provide the right information.

Group Work

Group work is one of the best methods of ensuring active learning or a facilitated classroom. (See Figure 2.9.) Actually forming the groups can be time-consuming, so do this as quickly as possible to not waste valuable time. If teaching a one-day course, keep the same group intact. If the course is over many days or weeks, students may need to change groups to gain different perspectives.

Groups can be formed in a number of creative ways. Randomly assign students to groups, have the students self-select their groups, or have the students count off. A number of other ways can also be used to divide the students into groups. Typically,

FIGURE 2.9 ◆ Group work can be accomplished in many forms and fashions. In the situation shown, students were involved in a communication exercise. The exercise was not directly related to the fire service; however, the activity emphasized the importance of effective communication.

the more control the instructor allows the students to have in selecting their group, the happier the student will be. The instructor will also find that the groups will have varying skill levels; keep this in mind when grading the groups. If enough characteristics are known about the students, the instructor can separate students in groups based on each individual's skill levels or personalities. Almost everything instructors do will be dependent on what they want to accomplish.

Facilitating Discussions

Discussion is one of the best forms of participatory lecturing. This can be effective for topics involving opinions, review of concepts, and beginning or ending classroom sessions. It is also useful in post-incident analysis sessions, including training drills.

Process for Facilitating Discussion

First, try to get all the students involved in the discussion. If it's a large group, the instructor may want to divide students into small groups to discuss the topic. This typically gets all the students involved in the discussion; occasionally, though, inattentive students will need to be redirected back to the group. After the groups have finished their discussion, reconvene into the larger group and have each group highlight the points they discussed. The instructor will want to put time limits on the group discussions and again when groups present their report to the entire class. The instructor doesn't have to comment on each group's presentation, and it's always wise to paraphrase what the group or a student states to make sure you understand. Be sure to compliment a good comment and redirect an inaccurate or incorrect statement to the class for correction.

When differences of opinion surface, the instructor will need to mediate. Mediation is a balancing act: The instructor will want to keep the discussion going without interjecting as the authority, which could damage the momentum of the discussion. Encourage the students to back their statements with documentation. Students need to learn that they need to substantiate findings. This is part of the real world. At times the instructor will need to remind everyone to respect the differing opinions of others.

After the discussion is complete, the instructor will need to pull the ideas together. Then the instructor will need to summarize what was discussed in the groups. Finish the discussion by providing follow-up information for additional study.

Practical (Psychomotor) Sessions

Experiential (or practical) sessions help to make training real. Employing practical sessions also aids the transference of information. Repeating the practice instills the procedure into the student's mind, and transference occurs. Use such instructional tactics as role-playing, games, simulations, and problem-solving tasks. (See Figure 2.10.)

FIGURE 2.10 ◆ Practical exercises can be as elaborate as an instructor likes. In this exercise individuals performed to the same level they would at an actual hazmat incident. *(Photo courtesy of James Clarke, Estero Fire Rescue)*

Tips for Practical Sessions

Explain the objectives to the students so that they have a clear understanding of what to do. Explain the benefits of performing the skill and how it relates to what they will be doing in the real world.

Practical sessions don't work well with large groups, so divide the students into smaller groups. The groups need to be small enough so that everyone has an opportunity to participate. Involve those who aren't participating in the drill or practical exercise by having them record the activity or be peer evaluators.

One of the hurdles to overcome as an instructor is knowing the material. The students do not. Speak slowly when giving directions so that students can comprehend what they are to do. Begin by providing a brief overview of the activity they are to perform; then provide the specific information. This meets the learning styles of both global and analytic learners. If the activity involves new equipment, give directions before handing out the equipment so the students can concentrate on the directions. Alternately, let them check out the equipment for a minute or so before beginning to give the instructions for the evolution. This will allow the students the ability to visualize the process.

Modeling the way something should be done is probably one of the best techniques to get a point across. Many people can perform a skill after seeing someone do it. For consistency sake, this is best accomplished if done one time for the entire group, including any adjunct faculty. Consistency is a critical factor in teaching practical skills. Instructors don't want adjunct instructors telling the students to do it one way when they will be testing them another way. If the instructor is working in a setting where the students rotate through various skill stations, they may need a quick review when they come to the station. This should not take as long to demonstrate as the original demonstration—but establish a time limit for each of the practical stations, and stick to it. This will keep participants (instructors *and* students) from lingering, plus keep the activity moving along and prevent students from getting bored.

Students need to be challenged. Begin with a simple or rote exercise and build toward a critical-thinking situation. At the completion of each session, recap and critique the events. First let the team leader or person performing the skill give his/her opinion of what was done correctly and incorrectly and then allow other student participants to give feedback (team, peer evaluators, recorders, etc.). The instructor should give feedback at the end. This should be presented in a positive-negative-positive format: begin with specific positive statements, follow with constructive criticism, and end with positive statements.

Classroom Control Issues with Facilitation

Instructors can easily lose control in an environment with too much interaction. Students may perceive that they are doing the instructor's job because the students are participating more actively in their learning. In fact, the student is more responsible for his/her learning. Facilitated learning is still new to some instructors, as well, and it may be perceived by other instructors that the instructor is not doing his/her job. Once these instructors realize the benefits of a facilitated learning environment, they will be more willing to use the technique.

The instructor needs to ensure that the student groups stay on track and do not wander too far off the topic being discussed. This goes for both large and small group discussions. Another danger is frustrated students. Students working in a small group facilitated environment may get frustrated when faced with a problem or issue they can't answer or if they don't understand what the instructor wants them to do. The

FIGURE 2.11 ◆ A timer can be used in some settings.

instructor needs to listen in on each group to make sure the groups are on the right path. Some groups will try to dominate the instructor's time and get the instructor to do the work. Be aware of this trap and just offer guidance, refocusing the group on the right direction.

The other danger is the time issue. Getting the small groups back into the large group may prove difficult. But start on time whether all the students are back or not. Otherwise this reinforces that it's acceptable to be late—plus instructors may lose their credibility/authority, and the students will be late the next time, as well. Here are some tips for getting participants back to order and ready to continue class.

1. Use a kitchen timer, watch alarm, or laptop timer. (See Figure 2.11.)
2. Say, "Now hear this!" into the microphone.
3. Create a verbal wave—clap hands or have everyone repeat, "Time's up."
4. Play music.
5. Make a unique sound—a gavel, a bell, a dinner gong.
6. Designate a timekeeper for the breaks who calls students back to the room.

Facilitation provides many benefits and when used effectively can be a powerful means to educate the students in the classroom. The next section discusses Socratic seminars, which provide more in-depth discussion and hone even further the skill of facilitation.

SOCRATIC SEMINARS

What does *Socratic* mean? The term comes from Socrates (ca. 470–399 B.C.), a classical Greek philosopher who developed a theory of knowledge. Socrates believed that the surest way to attain reliable knowledge was through disciplined conversation. He called this method *dialectic*—the art or practice of examining opinions or ideas logically, often by questions and answers, so as to determine their validity.

Socrates would begin by discussing the obvious aspects of any problem. He believed that through the process of dialogue, where all parties were forced to clarify their ideas, the final outcome of the conversation would be a clear statement of meaning. The technique appears simple, but it is intensely rigorous. Socrates would feign ignorance about a subject and draw out from the other person his fullest possible knowledge about it. His assumption was that by progressively correcting incomplete or inaccurate notions, one could coax the truth out of anyone. The basis for this assumption was an individual's capacity for recognizing lurking contradictions. If the human mind was incapable of knowing something, Socrates wanted to demonstrate that, too. Some dialogues, therefore, end inconclusively.

A Socratic seminar is a method of understanding information by creating a dialectic in class as regards a specific text. In a Socratic seminar, participants seek deeper understanding of complex ideas through rigorously thoughtful dialogue rather than by memorizing bits of information.

The Text. Socratic seminar texts are chosen for their richness in ideas, issues, and values plus their ability to stimulate extended, thoughtful dialogue. A seminar text can be drawn from readings in literature, history, science, math, health, and philosophy or from works of art or music. A good text raises important questions in the participants' minds, questions for which there are no right or wrong answers. At the end of a successful Socratic seminar, participants often leave with more questions than they brought with them.

The Question. A Socratic seminar opens with a question either posed by the leader or solicited from participants as they acquire more experience in seminars. An opening question has no right answer; instead, it reflects a genuine curiosity on the part of the questioner. A good opening question leads participants back to the text as they speculate, evaluate, define, and clarify the issues involved. Responses to the opening question generate new questions from the leader and participants, resulting in new responses. Thus, the line of inquiry in a Socratic seminar evolves on the spot rather than being predetermined by the leader.

The Leader. In a Socratic seminar the leader plays a dual role of leader and participant. The seminar leader consciously demonstrates thinking that leads to a thoughtful exploration of the ideas in the text by keeping the discussion focused on the text, asking follow-up questions, helping participants clarify their positions when arguments become confused, and involving reluctant participants while restraining their more vocal peers.

As a seminar participant, the leader actively engages in the group's exploration of the text. Doing this effectively means knowing the text well enough to anticipate varied interpretations and recognizing important possibilities in each. The leader must also be patient enough to allow participants' understanding to evolve and be willing to help participants explore nontraditional insights and unexpected interpretations.

Assuming this dual role of leader and participant is easier if the opening question is one that truly engages the leader as well as the participants.

The Participants. In a Socratic seminar participants are responsible for the quality of the seminar. Good seminars occur when participants closely study the text in advance, listen actively, share their ideas and questions in response to the ideas and questions of others, and search for evidence in the text to support their ideas. When participants realize that the leader isn't looking for right answers but rather wants them to think out loud and exchange ideas openly, they discover the excitement of exploring important issues through shared inquiry. This excitement creates willing participants, eager to examine ideas in a rigorous, thoughtful manner.

Guidelines for Participants in a Socratic Seminar

1. Refer to the text when needed during the discussion. A seminar is not a test of memory or to learn a subject. Your goal is to understand the ideas, issues, and values reflected in the text.
2. It's OK to "pass" when asked to contribute.
3. Don't participate if you are not prepared. A seminar should not be a bull session.
4. Ask for clarification.
5. Stick to the point currently under discussion; make notes about ideas you want to discuss later.
6. Don't raise hands; take turns speaking.
7. Listen carefully.
8. Speak up so that all can hear you.
9. Talk to each other, not just to the leader or teacher.
10. Discuss ideas rather than each other's opinions.
11. You are responsible for the seminar, even if you don't know it or admit it.

Expectations of Participants in a Socratic Seminar

Did the participants
Speak loudly and clearly?
Cite reasons and evidence for their statements?
Use the text to find support?
Listen to others respectfully?
Stick with the subject?
Talk to each other, not just to the leader?
Paraphrase accurately?
Avoid inappropriate language (slang, technical terms, sloppy diction, etc.)?
Ask for help to clear up confusion?
Support each other?
Avoid hostile exchanges?
Question others in a civil manner?
Seem prepared?

What Is the Difference Between Dialogue and Debate?

Dialogue is collaborative: Multiple sides work toward shared understanding.
Debate is oppositional: Opposing sides try to prove each other wrong.
In dialogue, one listens to understand, to make meaning, and to find common ground.
In debate, one listens to find flaws, to spot differences, and to counter arguments.
Dialogue enlarges and possibly changes a participant's point of view.
Debate defends assumptions as truth.
Dialogue creates an open-minded attitude: an openness to being wrong and an openness to change.
Debate creates a close-minded attitude, a determination to be right.

In dialogue, one submits one's best thinking, expecting that other people's reflections will help improve it rather than threaten it.

In debate, one submits one's best thinking and defends it against challenge to show that it is right.

Dialogue calls for temporarily suspending one's beliefs.

Debate calls for investing wholeheartedly in one's beliefs.

In dialogue, one searches for strengths in all positions.

In debate, one searches for weaknesses in the other position.

Dialogue respects all the other participants and doesn't seek to alienate or offend.

Debate rebuts contrary positions and may belittle or deprecate other participants.

Dialogue assumes that many people have pieces of answers and that cooperation can lead to a greater understanding.

Debate assumes a single right answer that somebody already has.

Dialogue remains open-ended.

Debate demands a conclusion.

Dialogue Is Characterized by:

Suspending judgment.

Examining our own work without defensiveness.

Exposing our reasoning and looking for limits to it.

Communicating our underlying assumptions.

Exploring viewpoints more broadly and deeply.

Being open to disconfirming data.

Approaching someone who sees a problem differently not as an adversary, but as a colleague in common pursuit of better solution.

Sample Questions That Demonstrate Constructive Participation in Socratic Seminars:

Here is my view and how I arrived at it. How does it sound to you?

Do you see gaps in my reasoning?

Do you have different data?

Do you have different conclusions?

How did you arrive at your view?

Are you taking into account something different from what I have considered?

Activity

Develop a class session using the Socratic seminar method. Present it to the rest of the group.

◆ SUMMARY

The adult methodology of learning, or andragogy, utilizes different instructional strategies to keep students motivated in the classroom. Using facilitation and conducting Socratic seminars are two of the most effective teaching tools for adult learners. Instructors need to adapt their presentations to adult students.

Review Questions

1. Define andragogy.
2. Describe the characteristics of an adult learner.
3. Discuss how to motivate adult learners.
4. Identify one instructional strategy in each of the four teaching modes and describe how to adapt to the strategy in the classroom.
5. Describe a few examples of body language.
6. Describe how to conduct a facilitated class.

References

Kearsley, G. *Explorations in learning and instruction: The theory into practice database,* from *http://tip.psychology.org/index.html.*

National Highway Safety Transportation Administration. August 2002. *National guidelines for educating EMS instructors.*

Bibliography

Beyer, B. 1997. *Improving student thinking: A comprehensive approach.* Boston: Allyn and Bacon.

Burke, J. (Ed.). 1989. *Competency-based education and training.* New York: The Fahner Press.

Cherry, R. 1990. "Keeping the spark alive." *JEMS,* March, 62–65.

Cherry, R. A. 1998. *EMT teaching: A common sense approach.* Upper Saddle River, NJ: Brady-Prentice Hall.

Chickering, A. W. and Gamson, Z. F. (Eds.). 1991. *Applying the seven principles for good practice in undergraduate education: New directions for teaching and learning.* San Francisco: Jossey-Bass.

Dalton, A. 1996. "Enhancing critical thinking in paramedic continuing education." *Prehospital and Disaster Medicine, 11(4),* 246–53.

Entwistle, N. J. 1983. *Understanding student learning.* New York: Nichols Publishing.

Greive, D. 1991. *A handbook for adjunct/ part-time faculty and teachers of adults.* Cleveland, OH: INFO-TEC, Inc.

Hodell, C. 1997. "Basics of instructional systems development." *ASTD Info-line, 9706.*

Johnson, D., Johnson, R., and Smith, K. 1998. "Maximizing instruction through cooperative learning." *AAHE Prism, February.*

Johnson, D., Johnson, R., and Smith, K. (n.d.). *Active learning: Cooperation in the college classroom.* (2d ed.). Edina, MN: Interactive Book Company.

Kearsley, G. *Explorations in learning and instruction: The theory into practice database.* Retrieved January 14, 2004 from *http://tip.psychology.org/index.html.*

Knowles, M. 1975. *Self-directed learning.* Chicago: Follet.

Knowles, M. 1984. *The adult learner: A neglected species* (3d ed.). Houston, TX: Gulf Publishing.

Knowles, M. 1984. *Andragogy in action.* San Francisco: Jossey-Bass.

Lin, Y., and McKeachie, W. J. 1999. *College student intrinsic and/or extrinsic motivation and learning.* Washington, DC: American Psychological Association.

Self-talk for teachers and students: Metacognitive strategies for personal and classroom use. Boston: Allyn and Bacon.

MacGregor, J., Cooper, J., Smith, K., and Robinson, P. 2000. *From small groups to learning communities: Energize large classes. New directions for teaching and learning.* Indianapolis: Jossey-Bass.

McCarthy, B. 1987. *The 4MAT system*. Barrington, IL: Excel, Inc.

McClelland, D. C. 1987. *Human motivation*. New York: Cambridge University Press.

McClincy, W. 1995. *Instructional methods in emergency services*. New Jersey: Brady-Prentice Hall.

Novak, J. 1998. *Learning, creating using knowledge*. Hillsdale: Lawrence Erlbaum Associates.

Pike, R. 1994. *Motivating your trainees*. Minneapolis: Lakewood Publications.

Pike, R. 1994. *Creative training techniques handbook* (2d ed.). Minneapolis: Lakewood Books.

Socratic seminars retrieved January 2004.

Thiel, A., Stern, J., Kimball, J., and Hankin, N. 2003. *Trends and hazards in firefighter training*. (No. USFA-TR-100.) Emmitsburg, MD: Federal Emergency Management Agency.

Treffinger, D., Isaksen, S., and Dorval, K. 1996. *Climate for creativity and innovation: Educational implications*. Sarasota, FL: Center for Creative Learning.

Walsh, A., and Borkowski, S. 1999. "Mentoring in health administration: The critical link in executive development." *Journal of Healthcare Management, 44*(4).

Whittle, J. 2001. *911 Responding for life*. New York: Delmar Publishers.

Learning Theories

CHAPTER

Terminal Objective

The participant will be able to incorporate various learning theories and strategies into his/her classroom presentation skills and techniques.

Enabling Objectives

- Describe the process of learning.
- Describe and differentiate between the various learning theories.
- Identify and describe the various learning domains.
- Describe the various learning styles and how they affect learning in the classroom.
- Identify various learning disabilities and how to properly deal with them in the classroom.
- Identify the various generations and describe how they affect learning in the classroom.

JPR NFPA 1041—Instructor I

4-2.2 Assemble course materials given a specific topic so that the lesson plan, all materials, resources, and equipment needed to deliver the lesson are obtained.

4-3.2 Review instructional materials, given the materials for a specific topic, target audience and learning environment, so that elements of the lesson plan, learning environment, and resources that need adaptation are identified.

4-3.3 Adapt a prepared lesson plan, given course materials and an assignment, so that the needs of the student and the objectives of the lesson plan are achieved.

4-4.4 Adjust presentation, given a lesson plan and changing circumstances in the class environment, so that class continuity and the objectives or learning outcomes are achieved.

JPR NFPA 1041—Instructor II

5-3.2 Create a lesson plan, given a topic, audience characteristics, and a standard lesson plan format so that the job performance requirements for the topic are achieved, and the plan includes learning objectives, a lesson outline, course materials, instructional aids, and an evaluation plan.

5-3.3 Modify an existing lesson plan, given a topic, audience characteristics, and a lesson plan, so that the job performance requirements for the topic are achieved, and the plan includes learning objectives, a lesson outline, course materials, instructional aids, and an evaluation plan.

5-4.2 Conduct a class using a lesson plan that the instructor has prepared and that involves the utilization of multiple teaching methods and techniques, given a topic and a target audience, so that the lesson objectives are achieved.

JPR NFPA 1041—Instructor III

6-3.4 Modify an existing curriculum, given the curriculum, audience characteristics, learning objectives, instructional resources, and agency requirements, so that the curriculum meets the requirements of the agency, and the learning objectives are achieved.

◆ INTRODUCTION

An instructor needs to understand how the students in the class are going to learn. An instructor's teaching style in a lot of instances reflects his/her learning style, but if the students have differing learning characteristics than how the instructor presents the material, the student may have a difficult time comprehending the material easily. Let's look at the various learning styles. (See Figure 3.1.)

◆ LEARNING

Learning is defined in many different ways. Even though each of these definitions may define learning differently, there are certain characteristics that should be part of the definition. Learning should:

- Create a lasting change in behavior.
- Result from practice or experience.
- Impart the capacity to behave in a particular manner.
- Occur throughout life.
- Be an internal change that is measurable externally.

Three circumstances affect learning. The first is the person's previous experience. Are the prerequisite skills and knowledge in place so that learning can occur? Motivation to learn, or attitude, is the second circumstance. The degree to which a person wants to learn, or has the incentive to change his/her behavior, will affect the learning outcome. Adult learners typically want to participate for multiple reasons, and therefore, motivation is generally not a problem.

FIGURE 3.1 ◆ Learning can take place in many environments. It is usually associated with a classroom setting.

The third circumstance is having an appropriate instructional method in a setting that facilitates optimal learning. For example, if teaching hose loads, and the process is presented only in lecture format with no hands-on application, the instruction will likely be less effective than if the cognitive knowledge is followed by psychomotor skills practice. (See Figure 3.2.)

◆ LEARNING THEORIES

A number of learning theories will be discussed in this chapter: behaviorism, socialism, constructivism, cognitivism, and conditions of learning. An instructor needs to gain a basic understanding of each of these theories and how they relate to instruction and/or course design and development. The theories are adapted from the *Explorations in Learning & Instruction: The Theory into Practice Database* website.

BEHAVIORISM

Overview

Behaviorism states that learning has occurred when there are changes in the form or frequency of an observable behavior. The behaviorist approach is referred to as the "sage on the stage" because the expert instructor is the center of the learning experience. The roots of behaviorism can be traced back to Socrates and ancient philosophical methods.

Notable behaviorist Edward Thorndike's learning theory represents the original stimulus-response (S-R) framework of behavioral psychology that posits that there is a stimulus, and from the stimulus a response is elicited. Therefore, learning is the result of the individual forming an association between stimuli and responses. The nature and

Figure 3.2 ◆ After firefighters learn about a skill in a classroom, it is good to have them practice the skill.

frequency of the S-R pairings strengthens or weakens the associations or "habits." Trial and error learning is a paradigm for S-R theory. This is a result of certain responses dominating others due to the rewards. Behavioral theory was based on the theory that learning could be explained without referring to any unobservable states.

Thorndike's theory consists of three primary laws:

1. *Law of effect:* Responses to a situation that are followed by a rewarding state of affairs will be strengthened and become habitual responses to that situation. A corollary of the law of effect was that responses that reduce the likelihood of achieving a rewarding state (i.e., punishments, failures) will decrease in strength.
2. *Law of readiness:* A series of responses can be chained together to satisfy some goal that will result in annoyance if blocked.
3. *Law of exercise:* Connections become strengthened with practice and weakened when practice is discontinued.

The theory suggests that transfer of learning depends on the presence of identical elements in the original and new learning situations—that is, transfer is always specific, never general. In later versions of the theory, the concept of "belongingness" was introduced: Connections are more readily established if the person perceives that stimuli or responses go together.

Example

The classic example of Thorndike's S-R theory was a cat learning to escape from a "puzzle box" by pressing a lever inside the box. After much trial-and-error behavior, the cat learned to associate pressing the lever (S) with opening the door (R). This S-R connection was established because it resulted in a satisfying current need: escaping from the box. The law of exercise specifies that the connection was established because the S-R pairing occurred many times (the law of effect) and was rewarded (law of effect) as well as forming a single sequence (law of readiness).

Principles

1. Learning requires both practice and rewards (laws of effect/exercise).
2. A series of S-R connections can be chained together if they belong to the same action sequence (law of readiness).
3. Transfer of learning occurs because of previously encountered situations.
4. Intelligence is a function of the number of connections learned.

SOCIAL LEARNING THEORY

Overview

The social learning theory is associated with Alberta Bandura. The social learning theory emphasizes the importance of observing and modeling the behaviors, attitudes, and emotional reactions of others. Bandura, in his work *Social Learning Theory* (1977), states: "Learning would be exceedingly laborious, not to mention hazardous, if people had to rely solely on the effects of their own actions to inform them what to do. Fortunately, most human behavior is learned observationally through modeling: from observing others, one forms an idea of how new behaviors are performed; and on later occasions, this coded information serves as a guide for action." (See Figure 3.3.)

FIGURE 3.3 ◆ Firefighters should be prepared to encounter a variety of fires after practicing training evolutions during practical evolutions.

The social learning theory explains human behavior in terms of the interaction between cognitive, behavioral, and environmental influences. The component processes underlying observational learning are

1. Attention, including modeled events (distinctiveness, affective valence, complexity, prevalence, functional value) and observer characteristics (sensory capacities, arousal level, perceptual set, past reinforcement).
2. Retention, including symbolic coding, cognitive organization, symbolic rehearsal, motor rehearsal.
3. Motor reproduction, including physical capabilities, self-observation of reproduction, accuracy of feedback.
4. Motivation, including external, vicarious, and self-reinforcement.

Because it encompasses attention, memory and motivation, the social learning theory spans both cognitive and behavioral frameworks.

Example

The most common and pervasive examples of social learning situations are television commercials. Commercials suggest that drinking a certain beverage or using a particular shampoo will make a person popular and win the admiration of attractive people. Depending on the component processes involved, such as attention or motivation, a person may model the behavior shown in the commercial and buy the product being advertised.

Principles

1. The highest level of observational learning is achieved by first organizing and rehearsing the modeled behavior symbolically and then enacting it overtly.
2. Individuals are more likely to adopt a modeled behavior if it results in outcomes they value.
3. Individuals are more likely to adopt a modeled behavior if the model is similar to the observer and has admired status and the behavior has functional value.

CONSTRUCTIVIST THEORY

Overview

Constructivism is applied to both the learning theory and to epistemology—how people learn and the nature of knowledge. Its core ideas have been clearly expressed by John Dewey, among others, but there is widespread acceptance of this old set of ideas and new research in cognitive psychology to support it.

The term refers to the idea that learners construct knowledge for themselves; learners individually (and socially) construct meaning as they learn. Constructing meaning is learning; there is no other way.

Principles

1. Instruction must be concerned with the experiences and contexts that make the student willing and able to learn (readiness).
2. Instruction must be structured so that it can be easily grasped by the student (spiral organization).
3. Instruction should be designed to facilitate extrapolation and/or fill in the gaps (going beyond the information given).

COGNITIVE LEARNING THEORIES OF DISSONANCE AND COGNITIVE FLEXIBILITY

Two theories of learning related to cognitivism are cognitive dissonance and cognitive flexibility.

COGNITIVE DISSONANCE

Overview

According to cognitive dissonance theory, individuals tend to seek consistency among their cognitions (i.e., beliefs, opinions). When there is an inconsistency between attitudes or behaviors (dissonance), something must change to eliminate the dissonance. In the case of a discrepancy between attitudes and behaviors, it is most likely that the attitude will change to accommodate the behavior.

Two factors affect the strength of the dissonance: the number of dissonant beliefs and the importance attached to each belief. There are three ways to eliminate dissonance: (1) reduce the importance of the dissonant beliefs, (2) add more consonant beliefs that outweigh the dissonant beliefs, or (3) change the dissonant beliefs so that they are no longer inconsistent.

Dissonance occurs most often in situations where an individual must choose between two incompatible beliefs or actions. The greatest dissonance is created when the two alternatives are equally attractive. Furthermore, attitude change is more likely in the direction of less incentive, since this results in lower dissonance. In this respect dissonance theory is contradictory to most behavioral theories that predict greater attitude change with increased incentive (i.e., reinforcement).

Example

A training agency buys an expensive training device but discovers that it does not provide realistic training scenarios. Dissonance exists between the beliefs that the agency bought a good training device and that a training device should be realistic. Dissonance could be eliminated by deciding that the realism does not matter, since the training device is a small portion of the curriculum (reducing the importance of the dissonant belief) or by focusing on the training device's strengths such as safety (thereby adding more consonant beliefs). The dissonance could also be eliminated by getting rid of the training device, but this would be a costly alternative to changing beliefs.

Principles

1. Dissonance results when an individual must choose between attitudes and behaviors that are contradictory.
2. Dissonance can be eliminated by reducing the importance of the conflicting beliefs, acquiring new beliefs that change the balance, or removing the conflicting attitude or behavior.

COGNITIVE FLEXIBILITY

Overview

The cognitive flexibility theory of Rand J. Spiro, R. L. Coulson, and Paul J. Feltovitch focuses on the nature of learning in complex and ill-structured domains. The psychologists' theory is largely concerned with transfer of knowledge and skills beyond their initial learning situation. For this reason emphasis is placed on the presentation of information from multiple perspectives, and many case studies that present diverse examples are used. The theory also asserts that effective learning is context-dependent, so instruction needs to be very specific. In addition, the theory stresses the importance of constructed knowledge; learners must be given an opportunity to develop their own representations of information to properly learn.

Cognitive flexibility theory builds on other constructivist theories (e.g., Bruner, Ausubel, Piaget) and is related to the work of Salomon in terms of media and learning interaction.

Example

D. Jonassen, D. Ambruso, and J. Olesen (1992) describe an application of cognitive flexibility theory to the design of a hypertext program on transfusion medicine. The program provides a number of different clinical cases that students must diagnose and treat using various sources of information available (including advice from experts). The learning environment presents multiple perspectives on the content, is complex and ill defined, and emphasizes the construction of knowledge by the learner.

Principles

1. Learning activities must provide multiple representations of content.
2. Instructional materials should avoid oversimplifying the content domain and support context-dependent knowledge.
3. Instruction should be case-based and emphasize knowledge building, not transmission of information.
4. Knowledge sources should be highly interconnected rather than compartmentalized.

CONDITIONS OF LEARNING

Overview

This theory stipulates that there are several different types or levels of learning. The significance of these classifications is that each different type requires a different type of instruction. Psychologist Robert Gagne identifies five major categories of learning:

- Verbal information
- Intellectual skills
- Cognitive strategies
- Motor skills
- Attitudes

Different internal and external conditions are necessary for each type of learning. For example, for cognitive strategies to be learned, there must be a chance to develop new solutions to problems; to learn attitudes, the learner must be exposed to a credible role model or persuasive arguments.

Gagne suggests that learning tasks for intellectual skills can be organized in a hierarchy according to complexity: stimulus recognition, response generation, procedure following, use of terminology, discriminations, concept formation, rule application, and problem solving. This hierarchy identifies prerequisites that should be completed to facilitate learning at each level. Prerequisites are identified by doing a task analysis of a learning/training task. Learning hierarchies provide a basis for the sequencing of instruction. (See Figure 3.4.)

Figure 3.4 ◆ In this scenario firefighters practice throwing a rope rescue bag prior to doing evolutions involving rescuing individuals from a swift water exercise. *(Photo courtesy of James Clarke, Estero Fire Rescue)*

In addition, the theory outlines nine instructional events and corresponding cognitive processes:

1. Gaining attention (reception)
2. Informing learners of the objective (expectancy)
3. Stimulating recall of prior learning (retrieval)
4. Presenting the stimulus (selective perception)
5. Providing learning guidance (semantic encoding)
6. Eliciting performance (responding)
7. Providing feedback (reinforcement)
8. Assessing performance (retrieval)
9. Enhancing retention and transfer (generalization)

Example

The following example illustrates a teaching sequence corresponding to the nine instructional events for the objective, to recognize the various hose loads:

1. Gain attention—show variety of hose loads.
2. Identify objective—pose question: "What is a preconnect flat load?"
3. Recall prior learning—review definitions of hose loads.
4. Present stimulus—give definition of preconnect flat load.
5. Guide learning—show example of how to load preconnect flat load.

FIGURE 3.5 ◆ This photo could be shown to students to illustrate a flat hose load.

6. Elicit performance—ask student to perform the load.
7. Provide feedback—check load as correct or incorrect.
8. Assess performance—provide scores and remediation.
9. Enhance retention/transfer—show pictures of hose loads and ask students to identify a preconnected flat load (Figure 3.5).

Principles

1. Different instruction is required for different learning outcomes.
2. Events of learning operate on the learner in ways that constitute the conditions of learning.
3. The specific operations that constitute instructional events are different for each type of learning outcome.
4. Learning hierarchies define what intellectual skills are to be learned and a sequence of instruction.

◆ IDENTIFYING THE LEARNING DOMAINS

In Chapter 2 the term *andragogy* was discussed. Now pedagogy needs to be defined. *Pedagogy* is the art and science of teaching. The art of teaching involves creative aspects like instructional design, developing classroom presentation skills, and so on. The science of teaching is based in educational psychology and research and deals with learning theories and preferences, the domains of learning, and other aspects of learning.

The domains of learning are a tool for understanding how people think, feel, and act. By understanding the domains of learning, the instructor can better plan what needs to be taught and how far to go through the material. This is also referred to as the "depth and breadth" of the course.

Domains of Learning

In 1956, Benjamin Bloom and his colleagues in educational psychology developed a theory that assumes abilities can be measured along a continuum, and classified levels of thinking and learning behaviors. This taxonomy includes three domains of learning:

Cognitive (thinking)
Psychomotor (doing)
Affective (feeling)

The domains of learning are used in instructional design to write goals and objectives for a curriculum. This is discussed in more detail in Chapter 9. These also serve as a means for instructors to develop test questions, which is discussed more in Chapter 7.

Levels Within the Domains of Learning

As the student progresses from one level to the next within a given domain of learning, the degree of sophistication increases, and a deeper and fuller understanding of the material is required. The levels are classified as lower and higher levels. The first level or two of each domain is considered the lowest level. The levels beyond this are considered higher levels. Sometimes this strategy is confusing as there are no clear division points between high level and low level, resulting in a greater degree of subjectivity.

There are three categories for each of the domains: knowledge, application, and problem solving. Knowledge, the first and lowest level, helps students comprehend facts, procedures, and feelings. It includes simple skills or thought processes like imitation, recall, definitions of terms, receiving, and responding to new information.

The second level, of which some are low and some are high, is application. Building on the knowledge foundation, this level involves the integration and execution of principles, procedures, and values within specific situations. It includes precision in the skills execution, the application of principles, and valuing feelings and beliefs.

The third and highest level is problem solving. This level builds on application and indicates that mastery has been achieved. It involves the analysis of information, procedures, and feelings to modify and adapt specific tasks, depending on situations. When an individual has reached the farthest stage of this level, he/she is capable of metacognition. Metacognition is thinking about thinking.

Earlier in this chapter it was discussed that the language of the objective should clue an instructor in to the level of depth and breadth to cover for the material. Figure 3.6 shows a list of verbs commonly used to describe objectives for each domain of learning.

THE COGNITIVE DOMAIN

Dealing with didactic information, knowledge, and facts, the cognitive domain consists of six levels of sophistication from simplest to most complex. They are

Level 1: *Knowledge:* memorization and recall
Level 2: *Comprehension:* interpretation and understanding of the meaning behind the information
Level 3: *Application:* application of classroom information to real-life situations and experiences
Level 4: *Analysis:* separation of the whole into parts in order to analyze their meaning and understand their importance
Level 5: *Synthesis:* combining of pieces of information into a new or different whole
Level 6: *Evaluation:* making judgments and decisions about and with the information presented

These levels are discussed in more detail in Chapters 7 and 9.

Domain Verbs

Cognitive Domain

Knowledge: arrange, define, describe, identify, label, list, name, identify, match, memorize, order, recognize, recall, recite, repeat, know

Comprehension: classify, discuss, distinguish, explain, identify, indicate, locate, review, rewrite, summarize, tell, translate

Application: apply, choose, compute, demonstrate, operate, practice, prepare, solve

Analysis: analyze, calculate, compare, contrast, criticize, diagram, differentiate, distinguish, examine, experiment, evaluate, relate, separate, select

Synthesis: assemble, compose, construct, create, combine, design, formulate, organize, prepare, set up, summarize, tell, write

Evaluate: appraise, evaluate, judge, score

Psychomotor Domain

Imitation: repeat, mimic, follow

Manipulation: practice with minimal assistance, create, modify

Precision: perform without error, perform without assistance

Articulation: demonstrate proficiency, perform with confidence, perform with style or flair

Naturalization: perform automatically

Affective Domain

Receiving: accept, attempt, willing

Responding: challenge, select, support, visit

Valuing: choose, defend, display, offer

Organization: judge, volunteer, share, dispute

Characterization: consistently join, participate

FIGURE 3.6 ◆ Examples of verbs used in the cognitive, psychomotor, and affective domains

THE PSYCHOMOTOR DOMAIN

The psychomotor domain deals with skills, actions, and manual manipulation. It consists of five levels from basic to complex. They are

Level 1: *Imitation:* repeating the example given by instructor or role model
Level 2: *Manipulation:* practicing and creating one's own style
Level 3: *Precision:* performing skill without mistakes
Level 4: *Articulation:* proficient and competent performance of skill with style or flair
Level 5: *Naturalization:* mastery-level skill performance without cognition (sometimes referred to as "muscle memory" or automatic)

Fire fighting involves psychomotor skills, and the development of these skills is crucial to good tactics—among other outcomes they ensure the safety of team members. There are many ways to perform acceptable skill behaviors. A firefighter needs to know the steps of skills performance to effectively apply critical-thinking skills in field situations. The instructor needs to plan an approach to teaching students how to perform skills to maximize their abilities.

Five Levels of Psychomotor Skills

The first level of psychomotor skills is imitation. At this level—which can be summarized by the adage "See one, do one"—the student repeats what is done by the instructor. The instructor therefore needs to be careful to avoid modeling the wrong behavior for the students to emulate. Some skills are learned entirely by observation, with no need for formal instruction.

Level two is manipulation. Though *manipulation* sounds negative, it should not be construed as such in this context. At this level the instructor uses guidelines as a basis or foundation for the skill. An example would be skill sheets. Remember that students make mistakes—but making mistakes is a means to learn. When students make a mistake, they should be encouraged to think through the process and make corrective actions. Never mind "Practice makes perfect." No one will ever become perfect, but practice will hone the students' skills so that they'll be able to perform to almost perfection. Skills need to be performed correctly, and the instructor needs to ensure that the student is performing the fire fighting skills to the level of acceptable behavior to function as part of the team in the real world.

The next level is precision. At this level students have practiced sufficiently to perform a skill without mistakes, but generally they can only perform the skill in a limited setting. An example would be a student who can tie a bowline, but can't perform the skill with the same level of precision during an evolution.

The fourth level of psychomotor skills is articulation. At this level the student can integrate cognitive and affective components with skill performance. Students understand why the skill is done the way it is being instructed, they can identify when a certain skill is indicated, and they can perform the skill proficiently and in the appropriate context. In other words, the student can tie a bowline around an object and use it as it is designed to be used.

The final level of psychomotor skills is naturalization. At this level the student has reached the mastery level without cognition. Also referred to as "muscle memory," it is the ability to multitask effectively. The student can perform the skill perfectly during scenario, simulation, or an actual fire event. Be aware, though, that the new student, or the seasoned student learning a new skill, may not reach this level in the classroom.

Teaching Psychomotor Skills

When teaching psychomotor skills the whole-part-whole technique is useful. This requires the skill be demonstrated three times as follows:

- *Whole:* The instructor demonstrates the entire skill, beginning to end, while briefly naming each action or step.
- *Part:* The instructor demonstrates the skill again, step by step, explaining each part in detail.
- *Whole:* The instructor demonstrates the entire skill, beginning to end, without interruption and usually without commentary.

Demonstrating the skill using this repetition technique provides an accurate example of the skill. If students were not completely focused on the skill demonstration during one of the demonstrations, they have two other opportunities to watch the presentation and learn the procedure. This technique provides a rationale for how the skill has been performed. The instructor needs to decide whether the students may interject questions during the demonstration. Some instructors prefer to have the discussion and interaction during the second part. The great thing about doing the demonstration

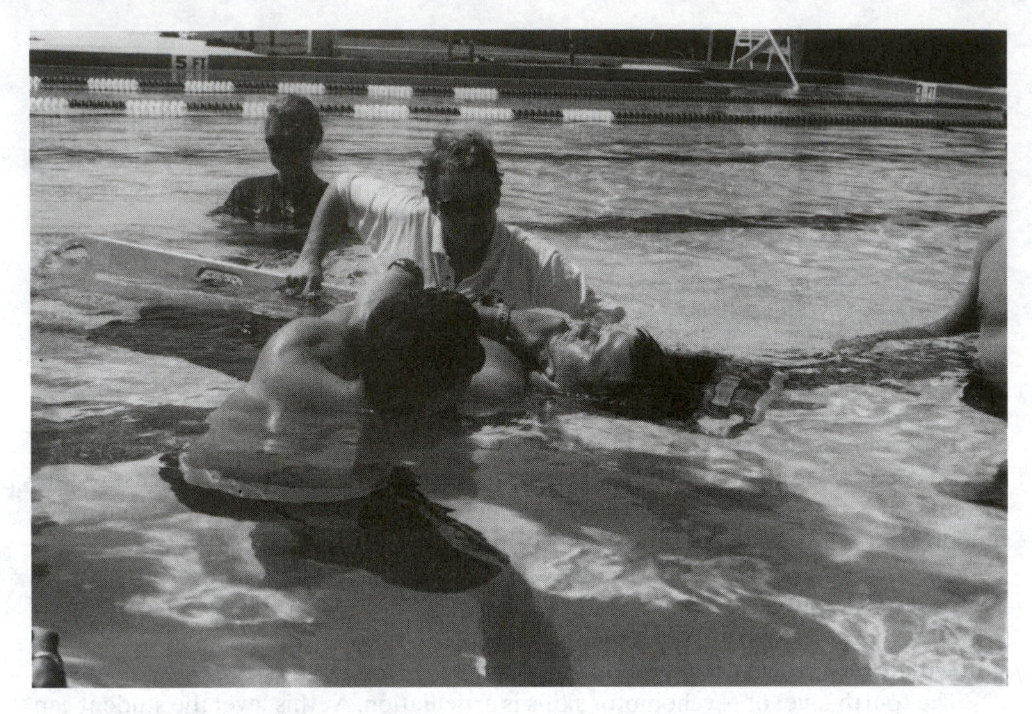

FIGURE 3.7 ◆ Instructors should demonstrate the skill they want the student to perform first.

in this manner is that it works well for both analytic and global learners. The analytic learners appreciate the step-by-step presentation, and global learners appreciate the overview. (See Figure 3.7.)

The Progress of Skill Acquisition

Students should be allowed to progress at their own pace. If the instructor pushes them too quickly, they may not understand what they are doing and won't acquire the good thinking skills that they need to perform the skills. The goal of teaching a psychomotor skill is to have the student be able to perform the entire skill from start to finish. However, the instructor should teach the skill by breaking it into individual parts, and let the student hone the technique. This requires the instructor to demonstrate the skill in its individual parts and then demonstrate the entire technique. The student should then be allowed to put the pieces together and perform the skill in its entirety.

The student needs to first master the skills and then apply them to a scenario or drill. They need enough time to practice and hone the skill in the scenario setting before being tested on it. As the student's learning progresses, the amount of direct instructor supervision should decrease.

The steps for learning the skill from the novice to the mastery level should follow this progression.

- First, the instructor demonstrates the skill to the student.
- Second, the student practices using a skill check sheet.
- Third, the student memorizes the steps of the skill until he/she can verbalize the sequence without error.
- Fourth, the student performs the skill stating each step as he/she performs it.
- Fifth, the student performs the skill while answering questions about his/her performance.
- Sixth, the student performs the skill in context of a scenario or actual situation.

More Than One Way

There are many ways to do things. Adult students need to be allowed to develop their own approach to a standard technique after they have mastered the skill. The instructor needs to focus on the acceptable behaviors that may not be the same as the instructor's way of performing the skill. If the end result is the same, and there were no safety infractions, encourage the student and give positive feedback. An instructor should spend time helping students develop high-level thinking skills so that they can differentiate between options and adequately solve problems.

Creating a Successful Skill Session

A number of things need to be accomplished to have a successful skill session. The first is to have all necessary equipment ready and set up before the session begins (Figure 3.8). Time is of the essence, and should not be wasted on finding and setting up the equipment. Second, the equipment needs to be realistic and current. The instructor doesn't want to be teaching skills with generation-old equipment so that when students get into the real setting, the equipment is different than what they were trained to use. Third, the equipment needs to be checked to make sure that it's in working condition. Nothing is more frustrating—and time consuming—for a student or an instructor than finding that equipment does not work. Lastly, if there are skill sheets, which is highly recommended, they should be standardized so all instructors are singing the same tune.

The skill sessions should provide ample time for the students to practice. It's very helpful to set up skill stations for students to practice at times other than designated skill sessions. And again, always model the skill behavior correctly. Always insist that the students respect equipment and skills and act accordingly.

It is essential that the instructor keep students active and involved in the skill session. Students who sit around waiting get very discouraged; they don't want to waste their time waiting to do the skill.

FIGURE 3.8 ◆ Equipment needs to be checked before using it whether you are in a training scenario or in the real environment. *(Photo courtesy of James Clarke, Estero Fire Rescue)*

Ensure that students are competent in their skills before using scenarios. When instructors do use scenarios, they need to be realistic. Any realism the instructor can add that makes the skill and the scenario more lifelike will increase the students' enthusiasm. Plus the students will tend to learn more from the session. Chapter 8 will discuss some of the realism that an instructor can add to a skill session.

THE AFFECTIVE DOMAIN

The affective domain is the development of judgment that determines how one will act. The affective domain deals with attitudes, beliefs, behaviors, emotions, and how much value an individual places on something. This is considered the most difficult domain to evaluate. Some of the words that describe the affective domain are

1. Defend
2. Appreciate
3. Value
4. Model
5. Tolerate
6. Respect

Many instructors believe the affective domain is one of the most difficult areas of thinking to influence. Indeed, some instructors believe that they *can't* influence students in this area. Instructors need to set aside their personal beliefs and emotions, and carefully cultivate the ethics and values of the fire profession. It is important that instructors understand the degree of responsibility they accept when they step into the classroom. The instructor has a strong influence on students. Students learn from instructors and emulate the behaviors that they model. Set a model for desired behavior in the classroom.

The Five Levels

The affective domain consists of five levels that range from simple to complex. They are

1. *Receiving* (Level 1): Awareness of the value or importance of learning the information and a willingness to learn
2. *Responding* (Level 1): Willingness to actively participate in the learning process and deriving satisfaction from doing so
3. *Valuing* (Level 2): Perception that behavior has worth
4. *Organization* (Level 3): Integration of different beliefs, reconciling differences
5. *Characterization* (Level 3): Development of one's own value system that governs one's behavior

The affective domain helps develop professional judgment. In turn, judgment often determines excellence. A student's ability determines his/her capability, and attitude determines the performance of the student. (The affective domain skills often make up the community's perception of the quality of service they receive.)

Some of the ideal characteristics of the affective domain include:

1. Kindness
2. Honesty
3. Compassion
4. Knowledgeability

Three levels of understanding within the affective domain were identified. Let's take a look at each of these levels.

The first level includes receiving and responding. Receiving is the awareness of the information or value the instructor is presenting. If students are willing to receive the information, they will pay attention and want to learn.

Responding is when the student does what is asked when required. The student can recall or recite correct answers according to what was taught in the classroom or read as part of an assignment. The student responds by doing the right thing, the right way when asked or when given other chances to respond.

The second level is valuing—the student is aware that a behavior has worth or value. A preference for a value shows that the student selects this behavior over others when given a choice. A commitment to a value means that the student always behaves this way and can defend or encourage this value in others.

The third level is organization and characterization. Organization integrates different beliefs based on experience. A student's good judgment comes from experience—though, ironically, this experience often develops out of bad judgment or poor decisions. This is why it's important to allow the student to make mistakes in the classroom and learn from the mistakes.

Characterization is behavior patterns that are so ingrained that they are part of the student's lifestyle—that given a number of situations involving the same value, the reaction will be automatic, consistent, and defensible. Characterization is when the person is so closely associated with the value that people may use the name of that value to describe the person.

The affective domain in the classroom can be described as the instructors being role models. They provide mentoring for students and set the example. Instructors need to be cognizant that students are constantly observing them. Even during breaks and outside of class the instructor is being observed and needs to set the example. The instructor is a role model. There will be times that instructors need to select adjunct instructors to assist them. Instructors need to carefully select their adjunct instructors to be sure that they model good values and meet the standards expected of an instructor.

The model values that you want your students to emulate are

1. Fairness
2. Compassion
3. Honesty
4. Punctuality
5. Dependability
6. Preparedness
7. Competence
8. Professionalism
9. Pride

Teaching Affective Domain Skills

A variety of presentation styles are appropriate to use for the affective domain. They include:

1. Case studies
2. Audio tapes of 911 calls
3. Discussion
4. Debate
5. Role-playing
6. Scenario

FIGURE 3.9 (A) AND (B) There are many ways to simulate an environment. In this scenario a master-stream from the engine is creating a simulated swift water scenario for personnel to practice rescuing victims. *(Photo courtesy of James Clarke, Estero Fire Rescue)*

EXAMPLES OF ACTIVITIES FOR EACH DOMAIN

- *Cognitive:* lecture, discussion, reading, diagramming, case studies, and drills
- *Psychomotor:* skills practice, scenarios, simulations, and role-playing
- *Affective:* modeling behaviors you expect the students to emulate (tolerance, punctuality, respect, kindness, honesty, and integrity), role-playing situations involving affective domain content, sensitivity training, and awareness courses

EVALUATION OF THE DOMAINS OF LEARNING

The domains of learning are often interdependent. Psychomotor skills development requires cognitive knowledge of the parts, concepts, and processes for practice to be most effective. (See Figure 3.9.)

For example: A student will achieve mastery of using a 1½-inch hose line to make an attack on a car fire faster if he/she can identify the needed equipment, understand the process of the skill, and recite the sequence of events to complete the skill before the skill is begun.

Some institutions or organizations encourage an environment where students do a high amount of experimenting as a means to learn, but even in these situations the students should be guided and mentored. These learning situations are most successful with those self-directed students who easily motivate themselves and have a passion for learning.

The instructor should review the course and lesson objectives to determine the depth and breadth of the course. It is best if the instructor can teach one level deeper than the objective requires because over time our memory loses retention of some of the information. (See Figure 3.10.)

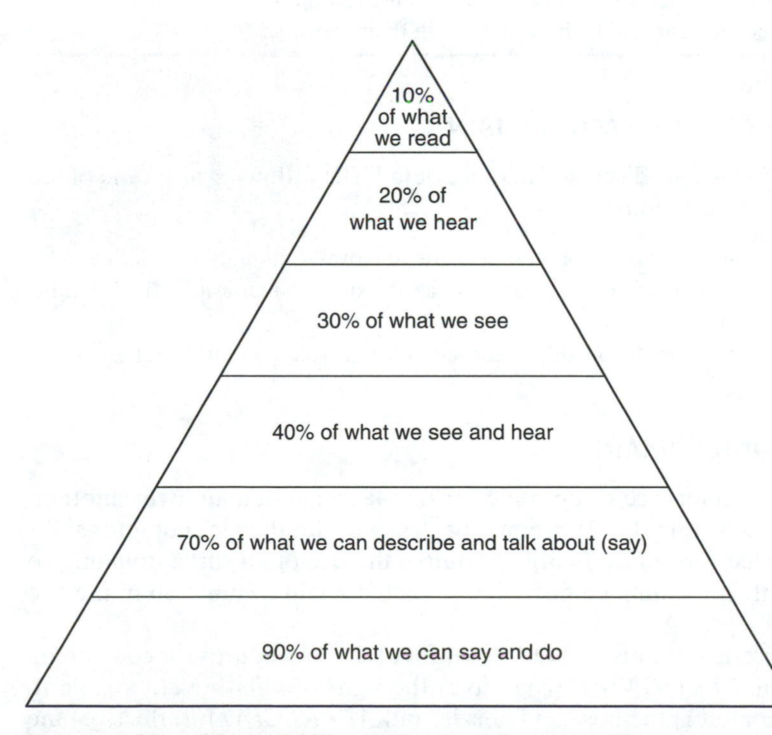

FIGURE 3.10 ◆ Pyramid illustrating learning retention

Research shows that the more senses that are engaged in the learning process, the more material is retained longer. Research also shows that the more times material is reviewed and reinforced, the more it is retained in long-term memory.

Let's take a look at some depth and breadth examples.

EXAMPLE 1

The first objective states that the student should take a list of the 12 parts of a pump panel and label those parts on a pump panel diagram; the second objective states that the student should draw a pump panel and label all the panel parts from memory. The first objective deals with a much lower level of cognition (knowledge) than the second objective (synthesis), so the instructor should be very thorough on teaching the second objective compared to the first.

EXAMPLE 2

The third objective states that the student should be able to take an empty SCBA cylinder and switch the regulator to a full tank. If the instructor has only discussed or demonstrated how to open the SCBA cylinder and check it for leaks, the students will probably fail an evaluation of this skill. (See Figure 3.11.)

EXAMPLE 3

The fourth objective states that the student should be able to list 10 different knots, but you only stressed 3 or 4 of them. The students won't likely successfully test on this objective unless they are highly self-motivated and learned the information on their own.

EVALUATION METHODS FOR EACH DOMAIN

In Chapter 7 evaluation will be discussed in more detail. The following are some of the evaluation methods for each domain:

Cognitive: written examinations, static presentations, and oral examinations

Psychomotor: skill competency exam, scenario-based exam, evaluation in field setting, on-the-job performance

Affective: class participation, leadership, peer supervision, role-modeling, adherence to policies

PREFERRED LEARNING DOMAIN

Most students have a preference or aptitude for one learning domain over another. Some students are excellent in the classroom but struggle with the psychomotor skills of fire fighting and vice versa. Fire fighting requires the use of all three domains, so students must meet the minimum competency in each domain to function in the fire fighting profession.

For example, a firefighter must *KNOW* (cognitive) the procedures for controlling a flashover condition, *RECOGNIZE* (cognitive) the signs of a flashover, be able to *ASSEMBLE* (psychomotor) the hose and nozzle, and *APPRECIATE* (affective) the level of severity that a flashover creates.

FIGURE 3.11 ◆ In this scenario the firefighter is at a higher level of cognition. He is actually using an SCBA in a simulated hazardous environment. *(Photo courtesy of James Clarke, Estero Fire Rescue)*

◆ LEARNING STYLES

HISTORY OF LEARNING STYLES

The variety of learning styles is not new. We can trace the history of learning styles back to Hippocrates, the Father of Medicine, who felt that the observable differences in human nature could be divided into four distinct groups called temperaments. His proposition was that there was a lack of balance in the secretions coming from the blood of the heart (sanguine), the bile of the liver (choleric), the phlegm from the lungs (phlegmatic), and the bile from the kidneys (melancholic). Others built on his theory, including the Greek physician and philosopher Galen (A.D. 129–200), who felt that temperamental differences could be seen in a positive light, and Paracelsus (1493–1541), a Swiss-born Renaissance healer who traveled Europe expanding his knowledge and building on Hippocrates' theory. Many others have continued to classify human nature in this way, adhering to the four groups, including psychiatrist Ernst Kretschmer in the 1920s.

Defining Learning Styles

R. M. Felder (1993) defines a student's learning style in part by answering these five questions:

- What type of information does the student preferentially perceive: Sensory (sights, sounds, physical sensations) or intuitive (memories, ideas, and insights)?
- Through which modality is sensory information most effectively perceived: visual (pictures, diagrams, graphs, demonstrations) or verbal (sounds, written and spoken words, and formulas)?
- With which organization of information is the student most comfortable: inductive (facts and observations are given, underlying principles are inferred) or deductive (principles are given, consequences and applications are deduced)?
- How does the student prefer to process information: actively (through engagement in physical activity or discussion) or reflectively (through introspection)?
- How does the student progress toward understanding: sequentially (in a logical progression of small incremental steps) or globally (in large jumps, holistically)?

Learning Style Inventories

Many learning style inventory instruments have been developed and are available to assess learning styles. We'll take a look at three of the leading inventories (Kolb, Myers-Briggs, and Gregorc).

The first learning style inventory model is the Kolb Learning Style. David A. Kolb developed the Learning Style Inventory, commonly called the LSI, as a means to evaluate the way people learn and work with ideas in day-to-day life. The Kolb Learning Style Inventory defines two preferred ways students learn information: abstractness or concreteness, and reflection or activity. It further defines the learning modes into learning styles by a representation of two of the four learning modes. The learning styles are classified as:

Type 1 (concrete, reflective)
Type 2 (abstract, reflective)
Type 3 (abstract, active)
Type 4 (concrete, active)

In relation to the preferred style for simulation learning, Kolb theorized that the converger is the learning style that tends to enjoy simulations more than the other learning styles.

The Myers-Briggs Type Indicator (MBTI) is probably the oldest learning style inventory in use today. It was originally published in 1923 and evolved out of Carl Jung's work on psychological types. The MBTI—which requires special training and certification to administer—classifies students into four groups: extraverts or introverts, sensors or intuitors, thinkers or feelers, judgers or perceivers.

The Gregorc Learning Style Inventory (Figure 3.12) is similar to the Kolb Learning Style inventory in that both break the learning into two types of preference, perception and ordering. This learning style delineator further breaks the two types of abilities into two qualities for each preference. The two qualities of perception are abstractness and concreteness. The two qualities of ordering are sequence and randomness. The preferences are combined into four types of learners. The four learning styles are concrete-sequential, concrete-random, abstract-sequential, and abstract-random.

A number of other learning style inventories that won't be discussed in this text may be worth considering. They include inventories such as Hemispheric Dominance Model,

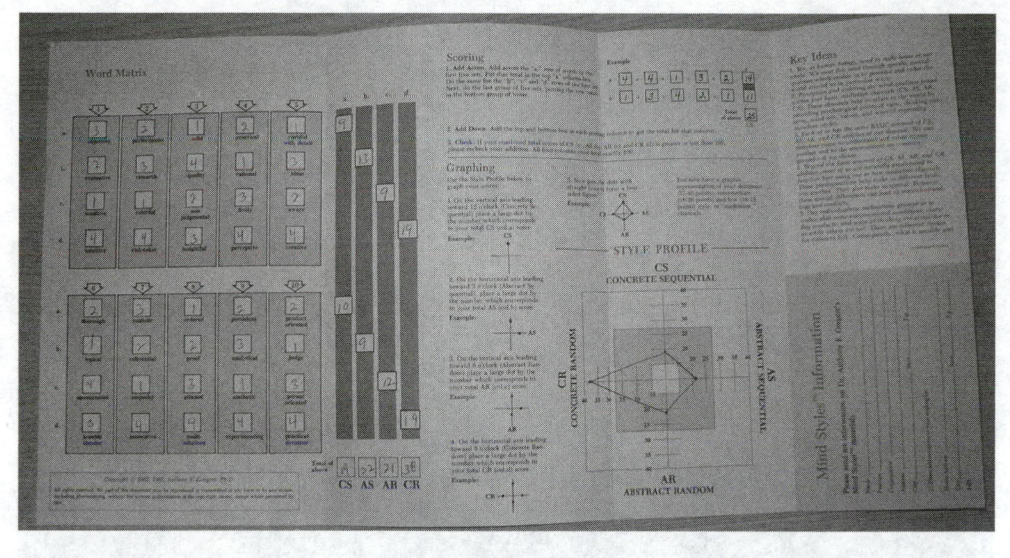

FIGURE 3.12 ◆ The Gregorc Mind Style Delineator is one learning style inventory that can be administered to individuals to determine their learning style. The Gregorc takes less than 3 minutes to complete.

Perceptual Modalities Model, Cognitive Styles Analysis, Developmental Cognitive Styles Metamodel, the Onion Model, Sternberg's Mental Self-Governmental Model, and Psycho-Geometric Personality Styles. The Hermann Brain Dominance Instrument and the Felder-Silverman Learning Style Model are used predominantly to assess learning styles in engineering students.

ASSESSING YOUR PERSONAL LEARNING STYLE

Instructors should know their learning preferences because instructors tend to teach the way they like to learn—though this may be a disservice to the students. When a misunderstanding arises in the classroom, the instructor should use personal knowledge of learning styles to reflect on how to present the material. It may be the reason for the misunderstanding.

LEARNING PREFERENCES

Here are a few examples of learning preferences, characteristics, and successful teaching techniques that can be incorporated into an instructor's teaching strategy:

- Auditory-visual-kinesthetic preferences
- Social and independent learning styles
- Analytic and global learning preferences

AUDITORY-VISUAL-KINESTHETIC LEARNERS

Auditory-visual-kinesthetic learners express a preference in the manner in which information is received.

Auditory Learners

Auditory learners learn best by hearing information, via discussion, listening, and verbalizing (Figure 3.13). These students do well when they record the audio portion of

Figure 3.13 ◆ Some individuals learn best by listening and are considered auditory learners.

Figure 3.14 ◆ Some learners learn best by visualizing. In this scenario, the instructor is showing the students the equipment. *(Photo courtesy of James Clarke, Estero Fire Rescue)*

FIGURE 3.15 ◆ There are many responsibilities that need to be performed during a training evolution. It is important that individuals work together to accomplish the goal of the training session. *(Photo courtesy of James Clarke, Estero Fire Rescue)*

the class and then listen to it again later. For this style of learning preference, use lectures, oral presentations, and class discussions to stimulate learning.

Visual Learners

Visual learners learn best by seeing information (Figure 3.14). Visual presentations of information include looking things up, writing things down, and "seeing" the words (forming word pictures in the brain). It's beneficial to provide handouts of content to these individuals. Some of the presentation styles for this group are videotapes, slide presentations, overheads, illustrations, posters, and other visual props.

Kinesthetic Learners

Kinesthetic learners learn best by physically manipulating information via handling and touching. The student benefits from taking things apart, making things work, using their hands, and tactile stimulation. The presentation style to use for this learning style includes three-dimensional models and replicas, laboratory sessions, scenarios, and role-playing. (See Figure 3.15.)

SOCIAL AND INDEPENDENT LEARNING STYLES

Social Learners

Social learners process information best when multitasking in busy environments with other people (Figure 3.16). Enjoying study sessions, group projects, and cooperative

FIGURE 3.16 ◆ It is important to be well organized when conducting a skill session. Make sure the students know what is expected of them before beginning the session. *(Photo courtesy of James Clarke, Estero Fire Rescue)*

learning, they work better by doing group work in class, participating in classroom discussions, study groups, and skills groups. This individual also does not mind music or other background noise.

Independent Learners

Independent learners process information best when working independently in quiet, undisturbed, regular study environments (Figure 3.17). The best way to reach these individuals is through reading assignments, written exams, papers, and reports.

ANALYTIC AND GLOBAL LEARNING THEORY

This theory describes the order in which a learner prefers to process information received: looking at the whole and then breaking it down into individual parts or looking at each individual part and then combining it into a whole. This is sometimes referred to as the right-brain and left-brain theory.

Global Learner (Right Brain)

Global learners need to first process the big picture, the overall view; then they can concentrate on the individual parts. Uncomfortable learning when they don't have a sense of the big picture, these students appreciate an overview of the material before an

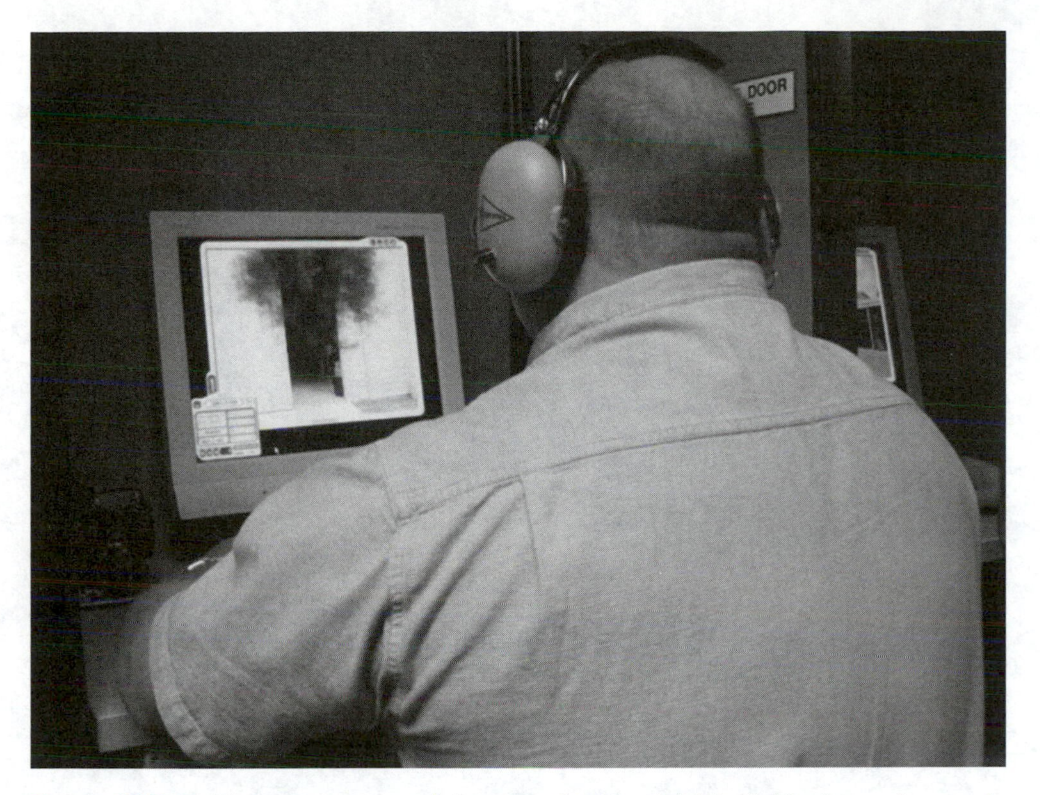

Figure 3.17 ◆ There are some training situations in which individuals need to work alone. *(Photo courtesy of Don Abbott, Phoenix Fire Department)*

instructor starts teaching. They process information globally and simultaneously and in images. They tend to be creative, artistic, imaginative, emotional, and intuitive and generally like working on teams. These individuals can best be taught using mental imagery, drawing, maps, metaphors, music and dance, and experiential learning.

Analytic Learner (Left Brain)

Analytic learners process information logically, sequentially, and in small parts. Uncomfortable with learning that occurs out of sequence, these students tend to enjoy spelling, numbers, thinking, reading, analysis, and speaking. This group can be best taught using lectures with outlines, reading assignments, and multiple choice exams.

The Forest or the Trees?

Do analytic and global learners see the forest or the trees?

Analytic learners separate the forest from the trees: They look at every tree in the forest before being comfortable enough to declare that they are in the forest.

Global learners will walk up to several trees, quickly declare it is a forest, and then will begin to look at the individual trees. (See Figure 3.18.)

FIGURE 3.18 ◆ You should make sure you see the whole picture, or in other words: make sure you see the whole forest, not just the trees.

THEORY OF MULTIPLE INTELLIGENCES

In establishing his theory of multiple intelligences, Howard Gardner determined that measuring IQ through a series of cognitive exercises does not fully measure the range of intelligences expressed by each individual. Gardner hypothesized that each person has aptitude in the following areas, with each individual having some areas with greater aptitude than others:

Linguistic: enjoys working with the spoken word and languages
Spatial: enjoys visual, artistic imagery, has the ability to construct visual pictures in his/her mind
Logical-mathematical: enjoys puzzles and problem solving requiring thought
Musical: enjoys music and understands the language of music
Body kinesthetic: has aptitude for sports and recreational activities involving bodily movements
Interpersonal: works well with others and is tuned into others who are around
Intrapersonal: enjoys self-reflection and introspection, is aware of his/her body

LEARNING DISABILITIES

The National Center for Learning Disabilities has defined some words commonly associated with learning disabilities:

Dyslexia—perhaps the most commonly known term, it is primarily used to describe difficulty with language processing and its impact on reading, writing, and spelling

Dyspraxia (apraxia)—difficulty with motor planning, impacts upon a person's ability to coordinate appropriate body movements

Dysgraphia—involves difficulty with writing; problems might be seen in the actual motor patterns used in writing; also characteristic are difficulties with spelling and the formulation of written composition

Auditory discrimination—a key component of efficient language use and necessary to "break the code" for reading; it permits being able to perceive the differences between speech sounds and to sequence these sounds into meaningful words

Visual perception—critical to the reading and writing processes as it addresses the ability to notice important details and assign meaning to what is seen

Attention Deficit (Hyperactivity) Disorder (ADD/ADHD)—may cooccur with learning disabilities; features can include: marked overactivity, distractibility, and/or impulsivity that in turn can interfere with an individual's availability to benefit from instruction

The term *learning disability* was introduced in 1963 to describe the characteristics of a group of individuals of at least average intelligence who seemed less capable of school success and who had unexplained difficulties in acquiring basic social and academic skills. Professionals note that individuals do not have a learning disability when the learning problems and/or failures are due primarily to:

impaired vision
mental retardation
emotional difficulties
hearing loss
environmental factors
physical disabilities

It is important to note that learning disabilities affect children and adults, and they range from relatively mild to severe. According to the National Center for Learning Disabilities, the criteria used to delineate a learning disability include whether the individual:

- Has an average or above-average intelligence,
- Exhibits an unexpected discrepancy between potential and actual achievement, and
- Performs poorly because of difficulty in one or more of the following areas: listening, reading, speaking, reasoning, writing, or mathematical skills.

There is no known exact correlation or cause of learning disabilities. A variety of factors may contribute to their occurrence, including:

- *Heredity*. Learning disabilities tend to run in families. Similar difficulties have been discovered in the same family.
- *Problems during pregnancy and birth*. Illness or injury during or before birth may cause learning disabilities. Learning disabilities may also be caused by the use of drugs and alcohol during pregnancy. Other factors may include: premature or prolonged labor, lack of oxygen, or low birth weight.
- *Incidents after birth*. Head injuries, nutritional deprivation, poisonous substances (e.g., lead), and child abuse.

Being lifelong, learning disabilities can have a significant impact on one's life. They can affect education, employment, daily activities, and interpersonal relationships. The following are some of the more predominant characteristics of learning disabilities in adults.

- Reading or reading comprehension
- Math calculations, math language, and math concepts
- Social skills, or interpreting social cues
- Following a schedule, being on time, or meeting deadlines
- Reading or following maps
- Balancing a checkbook
- Following directions, especially on multistep tasks
- Writing, sentence structure, spelling, and organizing written work
- Telling or understanding jokes

Instructors need to be aware that the individual may be able to learn information presented in one way but not in others. Or the individual may be able to communicate things verbally but have difficulty writing ideas on paper. These individuals may find it difficult to memorize information.

Individuals with learning disabilities may also have difficulty with social skills, which can spill over into the classroom. Instructors need to be sensitive to these issues that include:

- Self-esteem
- Interpersonal relationships
- Workplace functioning
- Community participation

Bear in mind that it isn't the instructor's responsibility to identify those students with a learning disability. It *is* the instructor's responsibility to reasonably accommodate them and be sensitive to the student's needs. A variety of resources offer information regarding this issue. The National Center for Learning Disabilities is an excellent source to gain more information and knowledge about learning disabilities. The information can be found on the Web at www.ncld.org, or contact the center at 1-888-575-7373. For legal issues regarding learning disabilities, call the U.S. Department of Justice, ADA information line: 1-800-514-0301.

EDUCATING THE GENERATIONS

An interesting issue for an instructor is how to deal with different generations in the classroom. If educating or training a new recruit, an instructor will probably be dealing with the X and Y generation. If teaching courses to those currently in the fire service, an instructor can end up dealing with four different generations in the classroom—not only generation X and Y, but baby boomers and maybe even a traditionalist or two.

For starters, instructors should figure out which generation *they* belong to. The range of years for each generation depends on what source is used. The name of each generation may even be different depending on the source. Your age may in fact put you between two generations—so how do you know which one is most apt? Simple. If you fall between two generations or have traits of two generations, you are referred to as being on the "fringe" or the "cusp" of a generation. Don't feel fragmented by

Era or Generation	Year Range	Age Range
Futuristic	Born after 2003	
Millennial, Millennium, Y	1982–2003	Up to early 20s
Generation X, 13ers	1961/65–1975/81	Early 20s–early 40s
Hackers (cusp)	1955–1965	Late 30s–late 40s
Boomers, Baby Boomers	1945–1964	Early 40s–late 50s
Traditionalists, Silent Generation	1925/35–1942	Early 60s–middle 70s
GI, Depression, Seniors	1901–1924/35	Middle 70s

FIGURE 3.19 ◆ Generations and their age ranges

being between generations. Being on the fringe is actually not that bad. Common opinion is that these individuals make some of the best managers, teachers, and marketers. By being on a cusp you have the ability to understand both generations and have more ability to relate to both of the generations.

It is important to define the age ranges and categories for each generation and set a standard reference when discussing the various generations. Figure 3.19 shows the common terms and age ranges for each of the generations. Remember, the age range varies according to the source that is being referenced.

GI Generation, Depression, Seniors. It would be a big mistake to devalue the essence of this generation. A lot of lessons can be learned from this significant era and as the population ages, this group is still intricately involved. As educators in the fire service, our students should be instilled with the values of this generation. These individuals support our agencies through tax monies and donations and are part of the community the fire service is sworn to protect. This generation is slowly disappearing, but is still an important part of the fire society. (See Figure 3.20.)

When describing this group, the words that come to mind are patriotic, loyal, fiscally conservative, and having faith in institutions. A fire service instructor won't likely have any of these individuals in the classroom, but at least note the traits of this generation to the students. These individuals are the end customer—a number of the incidents the students will respond to will involve this generation—and they should at least have a basic understanding of this group.

Traditionalists, Silent Generation. The members of this generation are considered cautious, unadventurous, unimaginative, withdrawn, and silent. A person will sometimes hear this group referred to as the "generation without a cause." This group has also been stigmatized with being the start of the "divorce epidemic"—and these individuals tended to marry earlier than previous generations. Having been born during an era of insecurity bracketed by the Great Depression and World War II, this group tends to be more conservative or traditional. (See Figure 3.21.)

This group, too, can be described with such words as patriotic, loyal, fiscally conservative, and having faith in institutions. Though this group is now retiring from the fire service, many of these individuals still like to remain active and involved by volunteering with a local fire department or by teaching in the fire service. If these folks sit in the classroom, give consideration to their traits. Instructors need to be careful about promises they make: Promises are a great

FIGURE 3.20 ◆ A senior member of the community may still be involved in the fire service and be part of your training session. *(Photo courtesy of James Clarke, Estero Fire Rescue)*

motivating factor for this generation, and they will expect people to come through with what they promise.

Boomer, Baby Boomer. Unless you have been under a rock for the past generation, you've heard the terms *boomer* and *baby boomer*. At times, the world may seem to revolve around the boomers, because this generation has had a tremendous amount of attention—and for good reason. The boomer generation, of which virtually every member is currently of working age, is 78 million strong; by contrast, Generation X has only 48 million members. So there is good reason why society pays attention to the baby boomers. Boomers have become sensitive to being portrayed as aging and slowing down; in fact, most are very interested in staying fit and remaining active. As an educator—especially a younger educator—don't view those older as less capable of performing tasks than a Generation Xer. (See Figure 3.22.)

The boomers are considered to be idealistic, competitive, questioning of authority, and are the "Me" generation. Despite their counterculture leanings, boomers are still somewhat traditional in their values—the nurturing and raising of children was of utmost importance to this generation—and can be a challenge in the classroom when mixed with Generation X. This is based upon the differences in culture and traits of the two generations.

Generation X, 13ers. This generation is referred to as the generation without a childhood, the "latchkey kids." The theory is based on the number of mothers thrust

Figure 3.21 ◆ The traditionalist generation is still active in many agencies. *(Photo courtesy of James Clarke, Estero Fire Rescue*)*

Figure 3.22 ◆ The baby boom generation is an active part of the American fire service.

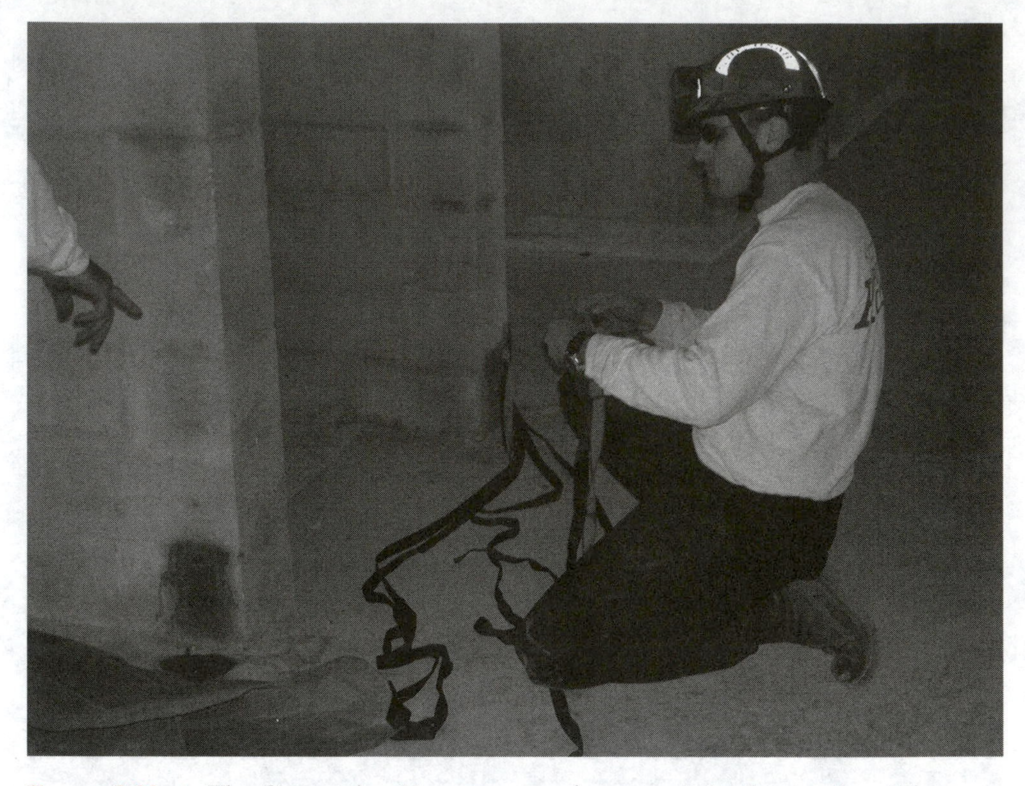

FIGURE 3.23 ◆ The Generation Xers are an active part of the fire service. *(Photo courtesy of James Clarke, Estero Fire Rescue)*

into the workplace, leaving children at home alone to fend for themselves. This generation has been confronted with drug addiction, AIDS, sexual freedom, uncontrollable violence, and environmental and other global problems created by past generations. The name comes from Douglas Coupland's 1991 novel *Generation X*. His label suggested an unknown quantity and emphasized the idea that Xers were more diverse and fragmented than any preceding generation.

In the workplace, members of Generation X often clash with baby boomers as they elbow the older generation out of the way. Education and training is the key to unlocking the relational existence between the two generations. (See Figure 3.23.)

Millennial Generation, Generation Y. The "Class of 2000" was born in this generation. The newest generation entering the fire service, the Millennial Generation or Generation Y, is being treated as a precious commodity. Parents and politicians alike are seeking a nurturing environment for this generation, hoping to shield them from the world of drug and alcohol abuse, violent behavior, AIDS, terrorism, and environmental disasters. Generation Y is considered the generation of hope. (See Figure 3.24.)

Futuristic Generation. This generation is just arriving, so why worry? You are absolutely correct to ask. It is because this generation will be the future firefighters—if that's what they will be called in the future. By the year 2050 the world's population is projected to increase from today's 5.5 billion to 11 billion, and the

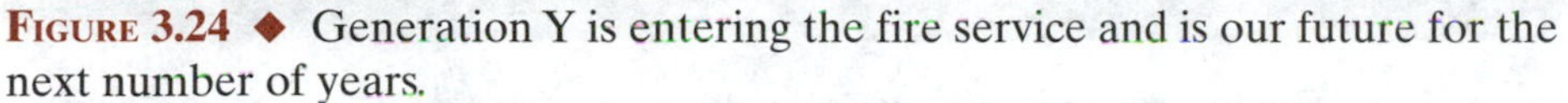

FIGURE 3.24 ◆ Generation Y is entering the fire service and is our future for the next number of years.

production of goods and service will have quadrupled. The world is rapidly changing; no doubt this generation will be viewing today's fire service as history—probably ancient history. This will be remarkable for the fire service. It has been said of traditional fire service that it's 100 years of tradition unimpeded by progress. Future generations won't maintain this approach, so as an instructor and member of the fire service, be ready to change. Instructors are training the new generations, the ones that will teach future generations. Future educators are being developed in the classes instructors conduct today. It is imperative that the classroom is where participants are eager to come and realize the value of what the instructor has to offer. (See Figure 3.25.)

TOWARD THE FUTURE

With Generation Xers and Yers putting a high priority on learning and developing new skills, instructors are challenged to continually raise the bar and bring new ideas to the classroom. The upcoming generations don't want to be bored with the same information. These generations also don't accept gray areas or unclear objectives. They want to know why they are "made to attend the class" and what is in it for them. Boomers see this as being disrespectful; however, Xers place a value on time and do not want to be wasting it on unnecessary things. How often have you asked why you do what you do? Does it really make sense to continue doing what always has been done, or can a change to make it better?

FIGURE 3.25 ◆ The future generation is still a number of years away from entering the fire service; however, now is the time to begin thinking about this generation and watch what will affect its members as they grow up.

The new generation is now entering the emergency service. Instructors are beginning to see these individuals in our classes. These individuals will provide even more unique situations for the instructors in the classroom. This group is considered to be self-reliant and stable, and appreciative of heritage, volunteerism, and escapism. They are also tolerant and accepting of diverse lifestyles and more than 40 percent must take remedial math or English after high school; vocational experience is limited; reading, writing, and emergency service technical knowledge is needed.

When it comes to training these generations, take into account a variety of issues. First, it is essential to accept them. These individuals are the future of the fire service. Do not alienate or neglect this group if the fire service is looking to train them as replacements when the current employees move on or retire. Plus the emergency service industry is beginning to feel the pinch of a tight labor market, especially with higher-paying jobs luring personnel into technology fields and away from emergency service jobs. Future emergency service personnel will disappear if actions are not taken now. (See Figure 3.26.)

Educators need to establish mentoring programs. Partnering older employees with newer employees lets each group learn more about one another. The newer generation wants hands-off management, but the instructor should still be there. This same mentality applies in the classroom. The instructor needs to allow this group to perform autonomously, and when there are questions, be there to answer them.

FIGURE 3.26 ◆ We need to remember to talk about the use of technology when we recruit the future generations of firefighters for the classroom.

Generations X and Y are independent learners. Having grown up in the technology era, this group is demanding more distance-learning classes. They expect changes to occur rapidly, whether in a traditional classroom or seated at computers to learn. A fire service instructor is challenged to make the class interactive with constant changes in how to deliver information. (See Figure 3.27.) Time was when the average attention span was thought to be about 50 minutes—or about how long one's posterior could endure sitting on a wooden chair in the classroom. Now, it's thought that change should occur at least every five or six minutes. This compressed time correlates with the change associated with current media devices and entertainment. Whether you like it or believe it, instruction has become another form of entertainment. With luck, the instructor adds the value of learning.

Generational Viewpoints

Traditionalists	"I learned it the hard way; you can, too!"
Baby boomers	"Train 'em too much and they'll leave."
Generation Xer	"The more they learn, the more they stay."
Generation Yer	"Continuous learning is a way of life."

Lancaster, L. C., and Stillman, D. (2002). *When generations collide.* New York, NY: HarperCollins.

FIGURE 3.27 ◆ Clashpoint around training

In her article "The Young and the Rest of Us," Jennifer Salopek outlines our lesson plan for the new instructional model:

- Make it make sense.
- Make it fun.
- Make it personal.
- Make it fast paced.
- Make it involving.
- Make it chunky.
- Make it safe to participate.
- Make it yours.
- Make it theirs.
- Make it learner centered.
- Make it positive.
- Make it matter.
- Make it comfortable.
- Make testing less stressful.
- Make trainees' experience count.
- Make it safe to disagree.

As the newer generations continue to enter the workforce and our classrooms, we need to deal with the ever-changing value systems—and adapt our delivery methods accordingly. (See Figure 3.28.)

Generational Trait Chart

Generation	Age	Influence	Traits
Millennium	0–early 20s	• Fall of the Berlin Wall • Expansion of technology • Mixed economy • Natural disasters • Violence • Drugs and Gangs	• Independence • Globally concerned • Health conscious • Cyberliterate
Generation X	Early 20s–early 40s	• *Sesame Street,* MTV • End of Cold War • Rise of personal computing • Divorce • AIDS, crack cocaine • Missing children on milk cartons and missing parents at home	• Techno savvy • Diverse • Independent • Skeptical • Entrepreneurial
Baby boomers	Early 40s–late 50s	• Booming birthrate • Economic prosperity • Expansion of suburbia • Vietnam, Watergate • Human rights movement • Sex, drugs, rock 'n' roll	• Idealistic • Competitive • Question authority • "Me" generation
Traditionalists	Early 60s–late 70s	• The Great Depression • The New Deal • World War II • The G.I. Bill	• Patriotic • Loyal • Fiscally conservative • Faith in institutions

FIGURE 3.28

◆ SUMMARY

The various aspects of learning include a number of learning theories that influence the education process. Learning styles can potentially influence both student and teacher. Learning disabilities can have a significant impact on a student's ability to succeed, and they need to be understood and addressed. And the generation gap plays out in the classroom as it does elsewhere in life, to good and ill affect.

Review Questions

1. Describe behaviorism learning theory and give an example.
2. Describe social learning theory and give an example.
3. Define a fire fighting skill using Gagne's nine steps of instructional events.
4. List the six levels of the cognitive domain.
5. List the five levels of the psychomotor domain.
6. List the six areas that describe the affective domain.
7. Describe how you would teach a fire fighting skill of choice using each of the domains.
8. Construct the retention pyramid.
9. Describe at least one learning style inventory.
10. Describe the differences between dyslexia, dyspraxia, and dysgraphia.

References

BridgeWorks retrieved on September 2002 from *http://www.generations.com.*

Cooper, S. S. 2001. *Learning styles.* Retrieved November 4, 2003, from *http://www.konnections.com/lifecircles/learningstyles.htm.*

Felder, R. M. 1996. "Matters of style." *ASEE, 6*(4), 18–23.

Felder, R. M. 1993. "Reaching the second tier: Learning and teaching styles in college science education." *Journal of College Science Teaching, 23*(5), 286–90.

Gregorc, A. F. 1982a. *An adult's guide to style.* Columbia, CT: Gregorc Associates, Inc.

Gregorc, A. F. 1982b. *Gregorc style delineator: Development, technical and administration manual.* Columbia, CT: Gregorc Associates, Inc.

Gregorc, A. F. 2003. *Frequently asked questions on style.* Retrieved November 16, 2003, from *http://www.gregorc.com/faq.html.*

Hedges, P. 1997. *Personality discovery: Personality patterns in teachers and their pupils.* Retrieved November 16, 2003, from *http://www.geocities.com/Athens/Aegean/9890/page5.html.*

Lancaster, L. C., and Stillman, D. 2002. *When generations collide.* New York, NY: HarperCollins.

Lowry-Mosley, M. G. 2003. *Running ahead: Style/learning inventories and Jung. The relationship between major contributors of style/learning inventories and Carl Jung's original theory of psychological types.* Retrieved November 4, 2003.

Mills, D. W. 2002. *Applying what we know to student learning styles.* Retrieved November 4, 2003.

National Highway Safety Transportation Administration. August 2002. *National guidelines for educating EMS instructors.*

McLoughlin, C. 1999. "The implications of the research literature on learning styles for the design of instructional material." *Australian Journal of Educational Technology, 15*(3), 222–41.

Ouellette, R. 2000. *Learning styles in adult education.* Retrieved November 4, 2003, from *http://polaris.umuc.edu/~rouellet/learnstyle/learnstyle.htm.*

Salopek, J. J. 2000. "The young and the rest of us." *Training and Development,* February, 26–30

Santo, S. 2003a. *Gregorc learning styles*. Retrieved November 4, 2003.

Santo, S. 2003b. *Kolb learning styles*. Retrieved November 4, 2003, from *http://www.usd.edu/~ssanto/kolb.html.*

Santo, S. 2003c. *Learning styles and personality*. Retrieved November 4, 2003, from *http://www.usd.edu/~ssanto/styles.html.*

Stahl, S. A. 1999. "Different strokes for different folks? A critique of learning." *American Educator* (Fall).

Bibliography

Adams, S. J. 2000. "Generation X." *Professional Safety,* January, 26–29.

Alch, M. L. 2000. "Get ready for the next generation." *Training and Development*, February, 32–34.

Anderson, C. W., and Krathwohl, D. R. (Eds.). 2001. *A taxonomy for learning, teaching and assessing: A review of Bloom's taxonomy of educational objectives*. Boston: Addison Wesley Longman, Inc.

Ausubel, D., Novak, J. D. and Hanesian, H. 1978. *Educational psychology: A cognitive view*. New York: Holt, Rinehart, & Winston.

Bandura, A. 1969. *Principles of behavior modification*. New York: Holt, Rinehart & Winston.

Bandura, A. 1973. *Aggression: A social learning analysis*. Englewood Cliffs, NJ: Prentice Hall.

Bandura, A. 1977. *Social learning theory*. New York: General Learning Press.

Bandura, A. 1986. *Social foundations of thought and action*. Englewood Cliffs, NJ: Prentice Hall.

Bandura, A. 1997. *Self-efficacy: The exercise of control*. New York: W. H. Freeman.

Bandura, A. and Walters, R. 1963. *Social learning and personality development*. New York: Holt, Rinehart & Winston.

Bloom, B. et al. 1956. *Taxonomy of educational objectives, book I: Cognitive domain*. New York: Longman.

Brehm, J. and Cohen, A. 1962. *Explorations in cognitive dissonance*. New York: Wiley.

Browne, M., and Keeley, S. 1998. *Asking the right questions: A guide to critical thinking* (5th ed.). Englewood Cliffs, NJ: Prentice Hall.

Bruner, J. 1960. *The process of education*. Cambridge, MA: Harvard University Press.

Bruner, J. 1966. *Toward a theory of instruction*. Cambridge, MA: Harvard University Press.

Bruner, J. 1973. *Going beyond the information given*. New York: Norton.

Bruner, J. 1983. *Child's talk: Learning to use language*. New York: Norton.

Bruner, J. 1986. *Actual minds, possible worlds*. Cambridge, MA: Harvard University Press.

Bruner, J. 1990. *Acts of meaning*. Cambridge, MA: Harvard University Press.

Bruner, J. 1996. *The culture of education*, Cambridge, MA: Harvard University Press.

Bruner, J., Goodnow, J., and Austin, A. 1956. *A study of thinking*. New York: Wiley.

Cherry, R. 1990. "Keeping the spark alive." *JEMS,* March, 62–65.

Cherry, R. A. 1998. *EMT teaching: A common sense approach*. Upper Saddle River, NJ: Brady-Prentice Hall.

Chickering, A. W. and Gamson, Z. F. (Eds.). 1991. *Applying the seven principles for good practice in undergraduate education: New directions for teaching and learning*. San Francisco: Jossey-Bass.

Dalton, A. 1996. "Enhancing critical thinking in paramedic continuing education." *Prehospital and Disaster Medicine*, *11*(4), 246–53.

Diestler, S. 1998. *Becoming a critical thinker* (2d ed.). Upper Saddle River, NJ: Prentice Hall.

Dunn, R., and Griggs, S. A. (Eds.) 2000. *Practical approaches to using learning styles in higher education*. Westport, CT: Greenwood Publishing Group.

Eiss, A. F., and Harbeck, M. B. 1969. *Behavioral objectives in the affective domain*. Washington, DC: NEA Publication Sales.

Ennis, R. 1996. *Critical thinking*. Upper Saddle River, NJ: Prentice Hall.

Entwistle, N. J. 1983. *Understanding student learning*. New York: Nichols Publishing.

Festinger, L. 1957. *A theory of cognitive dissonance*. Stanford, CA: Stanford University Press.

Festinger, L. and Carlsmith, J. M. 1959. Cognitive consequences of forced compliance [Electronic version]. *Journal of Abnormal and Social Psychology, 58,* 203–10. Retrieved on January 12, 2004 from *http://psychclassics.yorku.ca/Festinger/.*

Gagne, R. 1962. "Military training and principles of learning." *American Psychologist, 17,* 263–76.

Gagne, R. 1985. *The conditions of learning* (4th ed.). New York: Holt, Rinehart & Winston.

Gagne, R. 1987. *Instructional technology foundations*. Hillsdale, NJ: Lawrence Erlbaum Assoc.

Gagne, R. M., and Briggs, L. J. 1979. *Principles of instructional design*. New York: Holt, Rinehart & Wintson.

Gagne, R., and Driscoll, M. 1988. *Essentials of learning for instruction* (2d ed.). Englewood Cliffs, NJ: Prentice Hall.

Gagne, R., Briggs, L., and Wager, W. 1992. *Principles of instructional design* (4th ed.). Fort Worth, TX: HBJ College Publishers.

"Generation Gap," retrieved on September 2002 from *http://library.thinkquest.org.*

Goleman, D. 1998. *Working with emotional intelligence*. New York: Bantam Books.

Greive, D. 1991. *A handbook for adjunct/part-time faculty and teachers of adults*. Cleveland, OH: INFO-TEC, Inc.

Halpern, D. 1996. *Thought and knowledge: An introduction to critical thinking*. Hillsdale: Lawrence Erlbaum.

Johnson, D., Johnson, R., and Smith, K. 1998. "Maximizing instruction through cooperative learning." *AAHE Prism,* February.

Johnson, D., Johnson, R., and Smith, K. (n.d.). *Active learning: Cooperation in the college classroom*. (2d ed.). Edina, MN: Interactive Book Company.

Jonassen, D., Ambruso, D., and Olesen, J. 1992. "Designing hypertext on transfusion medicine using cognitive flexibility theory." *Journal of Educational Multimedia and Hypermedia, 1*(3), 309–22.

Kearsley, G. *Explorations in learning and instruction: The theory into practice database*. Retrieved January 14, 2004 from *http://tip.psychology.org/index.html.*

Knowles, M. 1975. *Self-directed learning*. Chicago: Follet.

Knowles, M. 1984. *The adult learner: A neglected species* (3d ed.). Houston, TX: Gulf Publishing.

Knowles, M. 1984. *Andragogy in action*. San Francisco: Jossey-Bass.

Kolb, D. A. 1984. *Experiential learning*. New York: Simon & Schuster Trade.

Langer, E. 1997. *The power of mindful learning*. Reading, MA: Addison-Wesley.

Lin, Y., and McKeachie, W. J. 1999. *College student intrinsic and/or extrinsic motivation and learning*. Washington, DC: American Psychological Association.

Self-talk for teachers and students: Metacognitive strategies for personal and classroom use. Boston: Allyn and Bacon.

MacGregor, J., Cooper, J., Smith, K., and Robinson, P. 2000. *From small groups to learning communities: Energize large classes. New directions for teaching and learning.* Indianapolis: Jossey-Bass.

McCarthy, B. 1987. *The 4MAT system*. Barrington, IL: Excel, Inc.

McClincy, W. 1995. *Instructional methods in emergency services*. New Jersey: Brady-Prentice Hall.

Miller, N., and Dollard, J. 1941. *Social learning and imitation*. New Haven, NJ: Yale University Press.

Millis, B., and Cottello, P. 1998. *Cooperative learning for higher education faculty*. Phoenix: Oryx Press.

National Center for Learning Disabilities, *http://www.ncld.org.*

Newble, D. I., and Entwistle, N. J. 1986. "Learning styles and approaches: Implications for medical education." *Medical Education 20,* 162–75.

Nix, D., and Spiro, R. (Eds.) *Cognition, education, and multimedia*. Hillsdale, NJ: Erlbaum.

Norman, G. R., and Schmidt, H. G. 1992. "The psychological basis of problem-based learning: A review of the evidence." *Academic Medicine 67* (9), 557–65.

Novak, J. 1998. *Learning, creating using knowledge*. Hillsdale: Lawrence Erlbaum Associates.

Parnes, S. 1997. *Optimize the magic of your mind*. Buffalo: Bearly Limited.

Pike, R. 1994. *Creative training techniques handbook* (2d ed.). Minneapolis: Lakewood Books.

Price, E. A. 1998. "Instructional systems design and the affective domain." *Educational Technology, 38* (6), 17–28.

Rideout, E. 2001. *Transforming nursing education though problem-based learning*. Sudbury: Jones and Bartlett Publishers.

Ruch, W. 2000. "How to keep gen X employees from becoming X-employees." *Training and Development,* April, 40–43

Spiro, R. J., Coulson, R. L., Feltovich, P. J., and Anderson, D. 1988. *Cognitive flexibility theory: Advanced knowledge acquisition in ill-structured domains*. In V. Patel (Ed.), Proceedings of the 10th Annual Conference of the Cognitive Science Society. Hillsdale, NJ: Erlbaum. Retrieved on September 15, 2003, from *http://www.ilt.columbia.edu/ilt/papers/Spiro.html.*

Spiro, R. J., Feltovich, P. J., Jacobson, M. J., and Coulson, R. L. 1992. "Cognitive flexibility, constructivism and hypertext: Random access instruction for advanced knowledge acquisition in ill-structured domains." In T. Duffy and D. Jonassen (Eds.), *Constructivism and the Technology of Instruction*. Hillsdale, NJ: Erlbaum.

Thiel, A., Stern, J., Kimball, J., and Hankin, N. 2003. *Trends and hazards in fire-fighter training*. (No. USFA-TR-100.) Emmitsburg, MD: Federal Emergency Management Agency.

Thorndike, E. 1913. *Educational psychology: The psychology of learning*. New York: Teachers College Press.

Thorndike, E. 1921. *The teacher's word book*. New York: Teachers College Press.

Thorndike, E. 1922. *The psychology of arithmetic*. New York: Macmillan.

Thorndike, E. 1932. *The fundamentals of learning*. New York: Teachers College Press.

Thorndike, E. et al. 1927. *The measurement of intelligence*. New York: Teachers College Press.

Thorndike, E. et al. 1928. *Adult learning*. New York: Macmillan.

Tishman, S., Perkins, D., and Jay, E. 1995. *The thinking classroom: Learning and thinking in a thinking culture*. Boston: Allyn and Bacon.

Treffinger, D. 1995. *Creativity, creative thinking, and critical thinking: In search of definitions*. Sarasota, FL: Center for Creative Learning.

Treffinger, D., Feldhusen, J., and Isaksen, S. 1996. Guidelines for selecting or developing materials to teach productive thinking. Sarasota, FL: Center for Creative Learning.

Treffinger, D., Isaksen, S., and Dorval, K. 1996. *Climate for creativity and innovation: Educational implications*. Sarasota, FL: Center for Creative Learning.

Walsh, A., and Borkowski, S. 1999. "Mentoring in health administration: The critical link in executive development." *Journal of Healthcare Management, 44*(4).

Whittle, J. 2001. *911 Responding for life*. New York: Delmar Publishers.

Wickland, R. and Brehm, J. 1976. *Perspectives on cognitive dissonance*. New York: Halsted Press.

Williams, L. 1983. *Teaching for the two-sided mind*. New York: Simon & Schuster.

Zywno, M. S. 2003. "*A contribution to validation of score meaning for Felder-Soloman's index of learning styles.*" Paper presented at the American Society for Engineering Education annual conference and exposition.

Learning Environments

Terminal Objective

The participant will be able to create a learning environment that is conducive for positive learning.

Enabling Objectives

- Design a classroom setting so that lighting, distractions, climate control or weather, noise control, seating, audiovisual equipment, teaching aids, and safety are considered.
- Diagram the various seating layouts for classroom instruction.
- List the advantages and disadvantages of each classroom layout.
- Adjust to differences in learning styles, abilities, and behaviors, given the instructional environment, so that lesson objectives are accomplished, disruptive behavior is addressed, and a safe learning environment is maintained.
- Describe the various cultures including ethnicity, religion, and age as they pertain to the classroom environment.
- List the components of a syllabus.
- Construct a syllabus.

JPR NFPA 1041—Instructor I

4-4.2 Organize the classroom, laboratory, or outdoor learning environment, given a facility and an assignment, so that lighting, distractions, climate control or weather, noise control, seating, audiovisual equipment, teaching aids, and safety, are considered.

4-4.5 Adjust to differences in learning styles, abilities, and behaviors, given the instructional environment, so that lesson objectives are accomplished, disruptive behavior is addressed, and a safe learning environment is maintained.

INTRODUCTION

The learning environment is the platform that the instructor creates to help students learn—and it can be one of the most daunting and exhausting aspects of teaching. It is difficult to make everyone happy when it comes to creating the perfect classroom; often there are elements that the instructor may not have any control over. (See Figure 4.1.)

The learning environment includes several important elements. First is the classroom design or layout. Accompanying the overall design are the physical characteristics, including temperature, seating arrangement, noise distractions, and acoustics. In addition to the physical characteristics of a classroom, it is imperative to create a learning environment that is harassment free. This chapter looks at each of these areas in an attempt to help build the classroom of optimal learning.

DESIGNING A CLASSROOM

In many instances the instructor is also the designer of the classroom. Ideally, the instructor can physically design a classroom. (See Figures 4.2 and 4.3.) In any regard, the classroom should be clean, safe, and well maintained to promote learning. The classroom must also comply with all federal, state, and local access laws regarding handicap accessibility. While this may not be the instructor's responsibility, it *is* the instructor's responsibility to make this information known to the proper authority.

FIGURE 4.1 ◆ The classroom environment can take many shapes and forms. This photo depicts the typical classroom that comes to most individuals' minds.

FIGURE 4.2 ◆ The Tennessee Fire Academy has a number of small classrooms used for arson labs.

FIGURE 4.3 ◆ When designing a classroom it is important to take many factors into account. This is an outdoor classroom with an erasable board for instructional use and misting fans for environment control.

PHYSICAL ELEMENTS

The classroom needs to be of adequate size to meet student needs and course requirements and be equipped with the appropriate tables, chairs, desks, and other necessary fixtures. If the instructor is responsible for actually purchasing the classroom furniture, it is important to make sure it is comfortable. If instructors are doing practical skills, they need to make sure there is adequate space to conduct the training as well as the proper equipment.

Ideally, the classroom will be located close to a food service area, rest rooms, and a break area. On the first day of class, the instructor needs to identify these areas for the students. The instructor will also want to point out the location of the fire extinguishers and explain what to do should the fire alarm sound while students are in the building. Identify the closest exits. If smoking is permitted, identify where it is allowed.

The classroom should have good environmental controls. This would include temperature regulation, adequate lighting, and light-blocking capabilities. It is also important to be able to minimize distractions outside the classroom, which may require shutting a door or closing blinds or shades.

TEMPERATURE REGULATION

The temperature of the classroom is the most difficult factor to control and one that an instructor needs to be sensitive to. This will be tricky because instructors will typically be warmer than most students—partly because they are moving about the room and partly because regardless of how experienced they are, the stress of standing in front of a class will likely raise their body temperature. Observe the students. If it's too warm, they will become sleepy; if it's too cold, they might lose their concentration. The instructor may not be able to adjust the temperature to a level that is comfortable for everyone, so just try to accommodate the majority. Keep in mind that this is probably one of the most difficult elements of creating an ideal environment. When instructors review their evaluations, the temperature will most likely be the item with the largest spread of agreement or disagreement. If an overwhelming number of students respond that the room was too cold or too warm, the instructor will need to adjust the classroom thermostat accordingly for future classes.

The fire service instructor will also be working in outdoor environments. Obviously the climate of the area will dictate what conditions the instructor will be dealing with. It is important to provide shelter for students who are not involved in the practical evolution. In warm environments, it's vital to keep students hydrated and cool. In cold environments protecting the students from frostbite is critical. It is important that crews train in various environments, since they will be working in all weather scenarios; however, safety must always take precedence over training in extreme environments.

SEATING LAYOUTS

Unless instructors are the only ones using the classroom and they are 100 percent certain the room is set up exactly as they want, they should arrive early enough to make sure the layout meets the needs. Classrooms can be set up in a variety of ways for best effectiveness. Ask a few questions to determine the ideal layout for the class. Should the instructor use tables? How should the chairs be arranged? Where is the best place to stand? Let's take a look at some of the arrangements that can be utilized depending on the class goals.

FIGURE 4.4 ◆ Most individuals think of the typical classroom style setting when designing a classroom. The desks and chairs are in straight rows for this type of layout.

Schoolroom. This layout is a typical classroom setting with all chairs and tables aligned in rows facing front. (See Figure 4.4.) This arrangement may not be the best design for a classroom full of adults, since it may remind them of their school years—and many adults may not have had a positive school experience!

This layout may also make it difficult for participants to concentrate—the student's attention may be diverted by the person ahead of them. Plus the view of the front of the classroom may be obstructed by the person in front of them. This may be unavoidable due to class size or facility arrangements, so the instructor may need to adapt to these limitations accordingly.

Advantages

- Maximizes instructor control
- Fair instructor mobility
- Ability to see visual aids in the front of the room to a certain degree
- Maximizes space

Disadvantages

- May lose students in the back of the room
- Discourages participation
- Reduces interaction

Auditorium Style. Think of the seating arrangement of a movie theater. Newer movie theaters are designed with stadium seating; that is, the seats are elevated so everyone has an unobstructed view. As such, this design provides a great environment for

FIGURE 4.5 ◆ An auditorium-style layout is ideal for lectures. It is easy for the instructor to see every participant and for students to see the instructor and media in the front of the classroom.

lectures. (See Figure 4.5.) The instructor won't find this to be the normal class environment, since the cost of this type of facility makes it unlikely that fire agencies will construct an auditorium; however, local schools and educational facilities may have these rooms available, and the instructor may want to consult with them to use these facilities for special courses. A downside to this setting is that it makes group interaction and/or small breakout sessions very difficult to accomplish.

Advantages

- Maximizes instructor control
- Good view of the front of the room
- Maximizes space

Disadvantages

- May lose students in the back of the room
- Discourages participation
- Reduces interaction

Bingo- or Cafeteria-Style. The name says it all. Tables are positioned in a row, and students face one another across the table (Figure 4.6). While this arrangement works great for meals or bingo, it is poor as a classroom layout, since it is difficult for

FIGURE 4.6 ◆ The cafeteria-style layout

the students to see the instructor. Plus students have to turn away from their notes or books to see any visual aids.

Advantages

- Fair instructor mobility
- Maximizes space

Disadvantages

- May lose people in the back of the room
- Discourages participation
- It is hard to simultaneously see visuals and write
- Reduces interaction
- Students may have trouble seeing the instructor

U-shape. Most people have been in a class with this type of room layout (Figure 4.7). It allows students to sit at their tables with everyone facing the open end of the U. Some instructors particularly favor this layout.

This arrangement allows for skill demonstrations and open discussions, but it is not useful for multimedia presentations because the instructor will likely block someone's view. This style allows for a more open environment; if there is small group interaction, this setup may be ideal. Be careful in using this class design for larger groups.

Advantages

- Excellent instructor mobility
- Encourages interaction among participants
- Encourages participation

FIGURE 4.7 ◆ A U-shaped classroom can have its advantages, but some students may have difficulty seeing media in the front of the classroom.

Disadvantages

- Students near the front may have to turn to see visuals.
- Uses more space per person
- May not work for large groups

King Arthur–Style. Wanting to promote a feeling of equality among his knights and to encourage freedom of speech, the legendary King Arthur created a round table, which eliminated a "head." This style allows for more interaction and is ideal for large group discussions. Also consider using a square configuration, especially if few visual presentations are to be used. (See Figure 4.8.)

Advantages

- Encourages interaction
- Encourages participation
- Maximizes writing space
- Excellent for teamwork in small or breakout groups

Disadvantages

- Poor instructor mobility
- Some students may have difficulty seeing visuals or the instructor.
- Uses more space per person
- Not suitable for large groups

Banquet Room. Think of the last banquet or wedding reception you attended. The tables are typically round and seat 8 to 12 individuals at each table. This arrangement

FIGURE 4.8 ◆ King Arthur–style denotes equality, but is not always a good layout for the classroom.

works well for a class designed for small group discussions. Otherwise this arrangement is poor for lectures or presentations. (See Figure 4.9.)

Advantages

- Encourages interaction
- Encourages participation
- Maximizes writing space
- Excellent for teamwork in small or breakout groups
- Good instructor mobility

Disadvantages

- Some students may have difficulty seeing visuals or the instructor.
- Uses more space per person
- Not suitable for large groups

V Formation. Think of the V formation as a flock of flying geese. Everyone has a good view. This style is great for lectures and also lends itself to group discussions and activities. It is easy to set up and doesn't take much additional effort to transpose a schoolroom layout to this style. (See Figure 4.10.)

FIGURE 4.9 ◆ The banquet-style layout is good for banquets and small group discussion, but it has a number of disadvantages for other class settings.

FIGURE 4.10 ◆ The V formation is considered an ideal layout for a classroom.

Advantages

- Excellent instructor mobility
- Excellent interaction among dyads or quads
- Encourages participation
- Provides writing space
- Good ability to see visual aids and the instructor
- Works well even with large groups

Disadvantages

- Poor for whole-group interaction

The instructor may want to experiment with other styles of classroom arrangements. Just keep in mind the students along with the goals for the class when designing the layout of the classroom seating. Don't be afraid to try something different—the instructor may just find the ideal classroom seating design.

NOISE DISTRACTIONS

Many distractions can hamper a student's ability to learn. Nowadays you'd be hard pressed to find someone who does not have a cell phone or pager, especially in the fire service. Ask the students to turn off their paging devices and cell phones or set them to vibrating alert so as not to distract others. Whenever possible, students should attend classes when off-duty to avoid interruptions.

A classroom's acoustics can also be an instructor's enemy. The instructor's voice projection is important to keep the class's attention. If the instructor is using a sound system, test the system prior to the start of class. Nothing is more frustrating than a poor sound system; but if the sound system is inadequate and the instructor failed to test it prior to class, it's a reflection of his/her ability to prepare.

Activity

Select a course topic you would like to teach, and describe the presentation format. Diagram the room layout, and explain why you selected this room configuration, giving the good and bad points for this setup.

CLASSROOM ATMOSPHERE

The classroom needs to be a safe and positive learning environment for both student and instructor—free from harm, discrimination, and harassment. The instructor needs to create an environment where students feel safe to exchange ideas and be creative in their problem solving. When students ask questions, they shouldn't be fearful of inviting emotional or physical discomfort. The instructor needs to be knowledgeable about the students and create a learning atmosphere diverse enough to meet the different learning styles of the students.

It is important to be an effective communicator. From the outset, the expectations need to be conveyed both in writing and verbally. If an instructor is teaching a course that will be conducted over multiple days, a course syllabus is a great mechanism to express expectations for the class. A syllabus is a simple contract between the instructor and the students. It should provide the students with what the instructor expects from them and what they will attain as a result of meeting the course expectations. A syllabus should include:

- Course participation
- Grading policy
- Attendance
- Assignments
- Instructor information
- Course materials
- Course schedule
- Course objectives
- Any other information that needs to be emphasized and communicated to the student.

Activity

Develop a syllabus for a 40-hour course that will meet once a week for 4 hours. Refer to other chapters in this text as a reference to construct the syllabus.

The material covered in the classroom needs to emphasize the most important aspects of the curriculum. In preparing to instruct the course, review the course objectives and use these guidelines to deliver the program. The job description and task analysis of the students is another means of gauging what is most relevant to cover in the classroom. If the class is a group of firefighters who work in an urban environment, they are probably not very interested in forest fire fighting. Likewise, a group of firefighters who serve a district with no building higher than three stories will most likely not be interested in high-rise fire fighting. Design the classroom to foster a positive environment and discourage negative behavior. In so doing, create an atmosphere where students learn more easily, and meet the needs of the student.

A good educational environment allows students to take risks. If the instructor is going to urge risk taking, the training must be designed so that the participants can put aside their normal responsibilities and see the information being disseminated from a different viewpoint. Students don't always feel comfortable performing in front of others—they may even feel uncomfortable or at risk when presenting skills in front of the instructor.

When an instructor creates an environment that allows students to perform in a safe arena, students will not only learn better, they will learn faster.

Additionally, when conducting skill sessions, remember that skills need to be repeated many times to ingrain the technique so that when students are called on to perform the skill in a real-life situation, they will do it without hesitation. Mastery of a subject is gained by the reinforcement of repetition.

Under the auspices of a facilitated learning environment, students are able to try new techniques and make mistakes. By letting them branch out, they will build confidence in what they are doing. Those students who have difficulty learning a skill should be allowed to practice until they have mastered the skill. A facilitated learning environment also permits those who have mastered the skill to work with others who have not.

STUDENT BEHAVIOR

Part of the learning environment is shaped by the behavior of the students in the classroom. To generate desirable student behavior, establish the group norms up front. Both desirable and undesirable behaviors create a learning environment.

Desirable behaviors or characteristics that need reinforcing include:

1. Integrity
2. Strong work ethic
3. Honesty
4. Courtesy
5. Respect
6. Engaged and active learning
7. Knowledgeability
8. Competence
9. Valuing lifelong learning

Undesirable student behaviors to discourage (or not reinforce) include the opposite behavior of each characteristic listed above as well as:

1. Lying
2. Cheating
3. Stealing
4. Violence
5. Intolerance
6. Prejudice
7. Carelessness
8. Unprofessional behavior and/or appearance

INFLUENCE OF THE INSTRUCTOR

The instructor has a significant impact on the learning environment. Indeed, in many instances instructors create the behavior in the classroom by their demeanor. Desirable behavior should, therefore, be modeled by the instructor. This can be referred to as the "mirror effect." What the instructor sees in the mirror is what the student sees. Now, some may not like looking into the mirror because they don't like a particular feature. In most cases this can't be changed. However, if a person dislikes something that he/she *can* change, change it! As the instructor you need to be able to look yourself straight in the mirror every day without any hesitation.

The second illustration of the mirror effect is that the instructor can use it to get someone to do something. Stated another way, actions speak louder than words. If instructors want someone to do something, the easiest way is to demonstrate what they want; and the student then emulates the actions. Instructors need to model the behavior that they want from their students. (See Figure 4.11.)

What are some desirable behaviors that an instructor should model? If instructors want the student to wear personal protective equipment (PPE) when operating on the training grounds, the instructors should be wearing their PPE. If instructors want students to arrive to class on time, they should arrive on time. These may sound simplistic, but things as simple as this can create a positive or negative learning environment by how well the instructor models them.

CULTURAL AWARENESS

The United States is a country of immigrants. This creates a diverse culture, and with this diversity comes a plethora of beliefs and traditions. Look at past immigration patterns. In 1940, 70 percent of immigrants were from Europe. By 1992, 37 percent were from Asia, 44 percent were from Latin America and the Caribbean, and only 15 percent were from Europe. More than 106 ethnic groups and 500 Native American tribes comprise the U.S. population. Many individuals who emigrate to the United States

FIGURE 4.11 ◆ The instructor needs to be the role model for students and maintain order in all environments. *(Photo courtesy of James Clarke, Estero Fire Rescue)*

want all the freedom and great lifestyle associated with being a citizen but don't wish to surrender their culture and its values. This results in a cultural pluralism—and it can also engender bias.

Bias

Overlooking cultural differences can be very difficult. In most cases, individuals do not even realize they have a cultural bias. Bias can come in all kinds of forms.

Age can be a difficult bias to overcome. In some cultures age is a huge barrier that may have an impact if the instructor is younger than the students in the classroom.

Trainer Tales

I started teaching early in my career. It was not always made evident, but many of the students in the class who had been in the fire service longer or were more mature would always have that look of "What can a young'un like you teach me?" I found that being confident in my approach and not trying to be someone I wasn't gave me more credibility. Although not everyone would listen, typically most would; and it was never a major problem. Be confident!

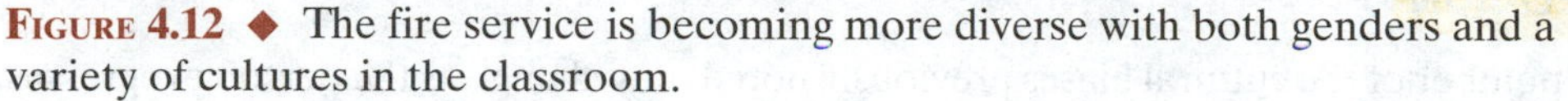

Figure 4.12 ◆ The fire service is becoming more diverse with both genders and a variety of cultures in the classroom.

Gender is another area to be aware of in the learning environment. (See Figure 4.12.) Female firefighters are still a minority group in the fire service. If the instructor is female, some will be biased against her as the instructor. If the instructor knows the material, gender should not matter. Another factor in gender is the different way men and women communicate. Women tend to communicate to form a relationship, whereas men often communicate to establish a hierarchy. It is also important when male and female students are in the class to be politically correct in word usage. This will help alleviate dissention or ill feelings toward the instructor or fellow students.

UNDERSTANDING ETHNICITY

To create a learning environment that embraces cultural differences, an instructor must first understand ethnicity. An ethnic background includes native language as well as cultural norms such as holiday observances, food preferences, social affiliations, and religious and health beliefs and preferences. While some Americans are comfortable with self-reliance and independence, this is not the case for all cultures.

Interdependence with relatives and friends varies among the cultures. In some cultures the family and the extended family are very important, and speaking out in a group that is not family may be uncomfortable. A classroom may not be seen as a safe environment for these people. Students may even be offended to the point that they feel shame and embarrassment may come to their family.

The way instructors present themselves and the atmosphere in the classroom may create an uncomfortable or unsafe environment for some cultures. Some cultures are nonaggressive and nonconfrontational. These students may not be comfortable making eye contact when conversing with a person in authority such as an instructor. In these instances the instructor needs to be understanding, and not feel resentment toward this person. Many people from other cultures address authority figures formally (by title or surname) until they receive permission to do otherwise. Additionally, gestures and speech patterns do not have universal meaning. A smile or a nod may be a sign of not understanding or not wishing to disagree with authority in some cultures. Snickering may be a sign of embarrassment and confusion for others. "Yes" may mean "I heard you" rather than "I agree with you." Some ethnic groups value silence as a sign of respect and attentiveness; for others it may be a sign of disagreement. Humor (particularly sexual in nature) and gestures are offensive to various cultures. There are courses and books available that cover cultural differences. It is good to get acquainted with at least the fundamentals of the various cultures that make up the community.

Activity

Select an ethnic background other than your own. Research the characteristics of this ethnicity, and prepare a report to present to your class.

RELIGION

A number of the cultural biases previously noted may change as the person progresses through his/her career in the fire service. Religion is one that may not. A student may not be available for class assignments due to a religious commitment or prayer times. Some religions don't consider Sunday as their holy day, as do most traditional Christians. For example, Seventh-Day Adventists and Jews consider Saturday their Sabbath. Muslims pray five times each day. The instructor needs to be respectful of religious differences.

Trainer Tales

I remember an instructor relaying a story about conducting a class in a Muslim community in a Northeastern state. At a certain point in the day, his students all stood, faced toward the east, and began to pray. After they had finished their prayers, the students returned to their seats and continued the class. For them, it was part of their day. For him, it was culture shock because he had not been prepared for it.

The Jehovah's Witnesses community forbids celebrations with the exception of the wedding anniversary. A student may not attend birthday, graduation, or holiday parties. Mormons fast for 24 hours once a month. Additionally, certain religious mandates impose specific dress codes that may conflict with practical evolutions or even practices within the fire service. The instructor will need to be upfront with students so they understand the job requirements. If their religious or cultural beliefs raise safety issues, they may not be able to be a firefighter. This is established by policy within each agency.

Activity

Research a religion other than your own, and present a report of the characteristics as it applies to training in the classroom.

OTHER INDIVIDUAL FACTORS

Many other factors may create an uncomfortable environment for various students. Married students may have other obligations, which might hinder or enhance their commitment. If a student is a parent, family issues can interfere with class responsibilities. Also, a student's income can limit access to education, transportation, and additional expenses—both during the student's initial training or afterward.

Personal habits may create an uncomfortable environment for the entire class. There are times the instructor may need to take a student aside and talk candidly to him/her about personal hygiene.

If instructing in a setting where students come from multiple agencies or multiple geographic locations, the instructor may need to pay attention to the interaction within the group. Some agencies may not get along, or being from a certain geographical location may be unacceptable to others in the classroom.

THE REALITIES

Culture is a very influential yet subtle power. Each of us is different, and each of us has values and beliefs. Instructors need to be aware of these different value and belief systems and be cautious not to exert their culture on the students. The fire service is a culture in and of itself. It is also a changing culture, and instructors need to be careful not to bias students before they go out into the real world. If a student is already working in the real world, the instructor needs to be aware of the cultures that surround that person to create an environment that is safe and one from which all can learn. Remember that it is difficult to know if an individual is offending someone. An instructor needs to be as respectful of others as possible.

◆ SUMMARY

The learning environment needs to be just that: conducive to learning. The students will rely on the instructor's ability to make the classroom an encouraging one to the best of his/her ability. Don't be afraid to think outside the paradigm and try something new. Education and training should be fun and exciting. Don't let the students end up saying it was any other way.

Review Questions

1. List the components of a syllabus.
2. Describe the student behaviors you should reinforce and those that you should not reinforce.
3. Select a religion, and describe the characteristics associated with it. Describe how you would handle a student from this religion in your classroom.

4. Describe some of the issues surrounding the classroom environment.
5. Select one classroom layout. Describe the advantages and disadvantages to using this layout and what type of class sessions would work best using this layout.

Bibliography

Adult and Continuing Education homepage. Retrieved on October 12, 2003 from *http://adulted.about.com/*.

Andrews, M. M., and Boyle, J. S. 1995. *Transcultural concepts in nursing care* (2d ed.). Boston: Scott, Foresman.

Bullock, K. A. 1997. "Shades of the rainbow." *Emergency Medical Services,* October, 28–33.

Davis, B., Wood, L., and Wilson, R. 1985. *A Berkeley compendium of suggestions for teaching with excellence*. Retrieved on September 28, 2003, from *http://teaching.berkeley.edu/compendium.*

Davis, B. 1993. *Tools for teaching*. San Francisco: Jossey-Bass.

Difficult Behaviors in the Classroom. (n.d.). Retrieved on November 11, 2003, from *http://www.hcc.hawaii.edu/intranet/committees/FacDevCom/guidebk/teachtip/behavior.htm.*

Enerson D. (Ed.)(n.d.). *Teaching at Chicago: A collection of readings and practical advice for beginning teachers*. Retrieved December 4, 2003, from *http://teaching.uchicago.edu/handbook.*

Galanti, G. 1997. *Caring for patients from different Cultures: Case studies from american hospitals* (2d ed.). Philadelphia: University of Pennsylvania Press.

Gardenswartz, L., and Rowe, A. 1998. *Managing diversity in health care*. San Francisco: Jossey-Bass.

Gardenswartz, L., and Rowe, A. 1999. *Managing diversity in health care manual: Proven tools and activities for leaders and trainers*. San Francisco: Jossey-Bass.

Gropper, R. C. 1996. *Culture and the clinical encounter: An intercultural sensitizer for the health professions*. Yarmouth: Intercultural Press.

Harrison, C. 1988. *Learning management*. ERIC Digests, 73. Retrieved on August 2, 2003, from *http://www.ed.gov/databases/ERIC_Digests/ed296121.html.*

Imel, S. 1994. *Guidelines for working with adult learners*. ERIC Digests, 154. Retrieved on January 10, 2004, from *http://www.ed.gov/databases/ERIC_Digests/ed377313.html.*

Imel, S. 1994. *Guidelines for working with adult learners*. ERIC Digests 77. Retrieved on January 10, 2004, from *http://www.ed.gov/databases/ERIC_Digests/ed299456.html.*

Imel, S. 1995. *Inclusive adult learning environments*. ERIC Digests 162. Retrieved on January 10, 2004, from *http://www.ed.gov/databases/ERIC_Digests/ed385779.html.*

Imel, S. 1995. *Teaching adults: Is it different?* ERIC Digests 82. Retrieved on January 10, 2004, from *http://www.ed.gov/databases/ERIC_Digests/ed305495.html.*

Leininger, M. 1978. *Concepts, theories, and practices*. New York: Wiley.

National Highway Safety Transportation Administration. August 2002. *National guidelines for educating EMS instructors*. Washington, DC.

O'Banion, T., et al. 1994. *Teaching and learning in the community college*. Washington, DC: American Association of Community Colleges.

Scholtes, P. 1988. *The team handbook*. Madison, WI: Joiner Associates.

Spector, R. E. 1996. *Cultural diversity in health and illness*. (4th ed.). Stamford, CT: Appleton and Lange.

Spector, R. E. 1996. *Guide to heritage assessment and health traditions*. Stamford, CT: Appleton and Lange.

Thiel, A., Stern, J., Kimball, J., and Hankin, N. 2003. *Trends and hazards in firefighter training*. (No. USFA-TR-100.) Emmitsburg, MD: Federal Emergency Management Agency.

Legal Issues in Instruction

Terminal Objective

The participant will be able establish a classroom environment that meets the legal ramifications as specified by local, state, and federal rules, regulations, and standards.

Enabling Objectives

- Discuss the NFPA's role in standards development in the fire service.
- List and relate the various NFPA standards that pertain to the fire service instructor.
- List and discuss the role of various local, state, and federal agencies as they pertain to the fire service instructor.
- Define negligence and describe how negligence affects the fire service instructor.
- Describe what constitutes harassment.
- Describe the various academic honesty issues.
- Explain confidentiality issues as they relate to the fire service instructor.
- Explain the affects of ADA and how this federal law impacts the fire service instructor.
- Explain copyright and describe the affects it has on the fire service instructor.

JPR NFPA 1041—Instructor I

4-2.3 Prepare training records and report forms, given policies and procedures and forms, so that required reports are accurately completed and submitted in accordance with the procedures.

4-5.4 Report test results, given a set of test answer sheets or skills checklists, a report form and policies and procedures for reporting, so that the results are accurately recorded, the forms are forwarded according to procedure, and unusual circumstances are reported.

JPR NFPA 1041—Instructor II

5-2.5 Coordinate training record keeping, given training forms, department policy, and training activity, so that all agency and legal requirements are met.

JPR NFPA 1041—Instructor III

6-2.2 Administer a training record system, given agency policy and type of training activity to be documented, so that the information captured is concise, meets all agency and legal requirements, and can be readily accessed.

6-2.3 Develop recommendations for policies to support the training program, given agency policies and procedures and the training program goals, so that the training and agency goals are achieved.

◆ INTRODUCTION

This chapter discusses various legal issues as related to the fire service instructor. Society today is more litigious than ever before (Figure 5.1). Fire service instructors and training institutions are not immune to lawsuits or liability for the safety and

Figure 5.1 ◆ Today's society has become very litigious. It is important that the fire service instructor take all precautions and exercise due regard when conducting training evolutions.

well-being of the students and instructors. Ignorance of the law is not an excuse and doesn't reduce responsibility. A fire service instructor should have a clear understanding of the common elements of the law as it pertains to the fire service and education.

It is the instructor's responsibility to inform the students of the applicable laws that pertain to them in the job they are about to do or are doing. This is one area in which most instructors are delinquent. It may be that many instructors don't like dealing with legal issues and are intimidated when it comes to teaching such matters. But no matter how hard instructors try they can't escape legal issues.

Instructors need to be aware of the laws that they are responsible for as instructors. As noted in the accompanying Trainer Tales, one can be criminally charged for not following the laws regarding live burns. The law is there for the instructor's protection and the students' protection. Nothing the instructor does is worth losing a life.

Trainer Tales

In September 2001, a firefighter died and two firefighters were injured while participating in a multiagency live-burn training evolution. Two of the firefighters were playing the role of firefighters in distress on the second floor. They became trapped for real when the training fire progressed up the stairwell, accelerated by a foam mattress that had been ignited on the first floor. Two of the victims, including the deceased, were found in the upstairs bedroom where they had been positioned for the exercise; a third firefighter jumped to safety from a rear bedroom on the second floor.

The firefighter who died had been a volunteer for only several weeks and had not received any formal fire suppression training before the live-burn evolution. Following the incident, investigators found numerous violations of NFPA 1403 including using personnel as victims, the absence of training in the case of the firefighter who died, the lack of charged hose lines prior to ignition, and the placement of training fires near egress routes. In addition, the department's first assistant chief was convicted of criminally negligent homicide for his actions leading up to and during the incident.

FIRE SERVICE TRAINING STANDARDS AND REGULATIONS

A variety of standards and regulations apply to fire service training programs. These standards are developed by the NFPA as well as local, state, and federal regulations.

THE NFPA

The National Fire Codes are a compilation of NFPA codes, standards, recommended practices, and guides developed through a consensus standards–development process. Representatives on these committees include such personnel as chief officers, volunteer representatives, union officials, and industry representatives. These codes are not legally binding per se, unless adopted, but they are the "standard of care" for the industry. When litigation is considered, lawyers often turn to the applicable standard of care in determining their course of action. It is up to decision-makers in political jurisdictions to determine levels of acceptable risk and the degree of liability exposure they will tolerate.

NFPA CODE DEVELOPMENT PROCESS

NFPA codes and standards are widely adopted because they are developed using an open, consensus-based process. All NFPA codes and standards are formulated and periodically reviewed by more than 5,000 committee members who are volunteers with a wide range of professional expertise. The volunteers serve on more than 200 technical committees and are overseen by the NFPA Standards Council. The NFPA Board of Directors appoints the 13-person Standards Council to administer the standards-making activities and regulations.

NFPA CODES AND STANDARDS

According to NFPA's "Regulations Governing Committee Projects," codes and standards are defined as follows:

- Code—a standard that is an extensive compilation of provisions covering broad subject matter or that is suitable for adoption into law independently of other codes and standards
- Standard—a document, the main text of which contains only mandatory provisions using the word "shall" to indicate requirements and which is in a form generally suitable for mandatory reference by another standard or code or for adoption into law. Nonmandatory provisions shall be located in an annex, footnote, or fine-print note and are not to be considered a part of the requirements of a standard.

NFPA 1500, Standard on Fire Department Occupational Safety and Health Programs

NFPA 1500 requires fire departments to establish and maintain a training and education program to help prevent occupational deaths, injuries, and illnesses. Furthermore, departments must provide training and education for the department members commensurate with the duties and functions they are expected to perform. Training must conform to the individual requirements for each fire department function (for example, firefighter, driver/operator, fire officer, and fire instructor).

In addition, NFPA 1500 requires live-fire evolutions to conform to NFPA 1403 and specifies that a qualified instructor must supervise all training and exercises.

NFPA 1403, Standard on Live-Fire Training

NFPA 1403 was developed after a training incident in 1982 that resulted in the deaths of two firefighters. Compliance with 1403 (in addition to improvements in personal protective equipment) has helped improve safety during live-fire training. Research indicates that since the implementation of this standard, major trauma or severe burn injuries during live-fire training have occurred mostly when a department has deviated from the standard. Some of the major points of NFPA 1403 include:

- Fuels such as pressure-treated wood, rubber, plastic, straw, and hay treated with pesticides or chemicals are prohibited.
- Flammable or combustible fuels are prohibited in acquired structures.
- A safety officer shall be designated.
- The maximum student-to-instructor ratio should be five to one.
- One instructor should be assigned to the backup line and to each designated sector assignment.
- EMS should be provided on site.

- No personnel shall play the role of a victim in the fire building.
- Only one fire at a time is permitted in acquired structures.
- Preevolution briefings must familiarize all personnel with the layout of the structure and location of emergency exits.

NFPA 1001, Standard on Firefighter Professional Qualifications; NFPA 1002, Standard for Fire Apparatus Driver/Operator Professional Qualifications; NFPA 1031, Standard for Professional Qualifications for Fire Inspector and Plan Examiner; and, NFPA 1021, Standard on Fire Officer Professional Qualifications

These four standards address the minimum training and knowledge required for personnel to operate at various levels of the fire service rank structure. These standards are designed to provide clear and concise job performance requirements to determine that an individual, when measured to the standard, possesses the skills and knowledge to perform a specific function.

NFPA 1041, Standard for Fire Service Instructor Professional Qualifications

NFPA 1041 identifies three distinct levels of responsibility for instructors. These qualifications were previously discussed in Chapter 1, but let's review. At Level I, the instructor must assemble necessary course equipment and supplies, review prepared instructional materials, and deliver instructional sessions based on the prepared materials. At Level II, an instructor is expected to be able to develop course materials independently and deliver those materials to students. Level III applies primarily to administrators and requires personnel to develop agency policy for training programs.

At all levels, NFPA 1041 requires instructors to be cognizant of the safety of their students to ensure that classroom and practical evolutions are conducted in a safe, controlled manner.

NFPA 1401, Recommended Practices for Fire Service Training Reports and Records

NFPA 1401 defines the practices for fire service organizations to establish, upgrade, or evaluate their training records and report systems to document clearly the performance and ability of individual and group activities as it relates to the following:

1. Compliance with personnel performance standards
2. Documentation of both internally and externally obtained career development training and education
3. Documentation for the purposes of certification and recertification
4. Cooperation with other agencies with which the organization executes joint specialty operations
5. Training required by regulatory and/or other agencies

NFPA 1410, Standard on Training for Initial Emergency Scene Operations

This standard contains the minimum requirements for evaluating training for initial fire suppression and rescue procedures used by fire department personnel engaged in emergency scene operations. It specifies basic evolutions that can be adapted to local conditions and serves as a standard mechanism for the evaluation of minimum acceptable performance during training for initial fire suppression and rescue activities.

NFPA 1451, Standard for a Fire Service Vehicle Operations Training Program

This standard contains the minimum requirements for a fire service vehicle operations training program. It outlines the development of a written training program, including organizational procedures for training, vehicle maintenance, and identifying equipment deficiencies, and for design, financing, and other areas. The knowledge and skills required of safety, training, maintenance, and administrative officers charged with developing and implementing the operations training program are also outlined within this standard.

LOCAL, STATE, AND FEDERAL REGULATIONS

Local, state, and federal regulations also govern training. State and local governments as well as individual fire departments establish minimum training requirements for fire suppression personnel. Such requirements typically—but don't necessarily—comply with NFPA standards. The federal government also plays a role in fire service training. For example, OSHA requires annual face-mask fit testing for firefighters. In Florida, State Statute 633.45 states that firefighters may only obtain required course instruction from approved instructors. Let's take a look at some agencies that have regulations governing fire fighting and training.

Environmental Protection Agency Standards

Local, state, and federal environmental regulations have a direct impact on conducting live-fire training evolutions. It is important to note that DEP 62–256.700(5)(A) requires that an agency follow NFPA 1403. Such regulations often limit the type and amount of fuel allowed to be burned and may require disposal of toxic products of live burns as hazardous waste. These restrictions limit how the instructor conducts live-fire training, ranging from burning in acquired structures to flammable liquid fire fighting. Individual as well as departmental liability, including severe fines, may result from violating such regulations. Additionally, the safety of training participants and staff may be compromised if the proper controls are not executed.

The implementation of various environmental regulations has created two choices for departments: (1) Adopt new, more environmentally friendly training methods or (2) Disregard the regulations. Other areas of environmental concern that should be considered when conducting training include air pollution regulations, runoff from hose streams, environmentally friendly foam, and the presence of asbestos or other obvious contaminants.

OSHA Regulations

The Occupational Health and Safety Act (OSHA) of 1970 regulates the employee's environment and practices to ensure the health and safety of our nation's workforce. These regulations are enforceable by law, and penalties may be applied for noncompliance. A number of states are OSHA compliant and have state enforcement offices. But even those states that are not OSHA compliant are not exempt from OSHA standards. An OSHA standard in many instances is considered the "standard of care." Therefore, agencies will typically be held to this level of accountability.

Regulation 29CFR1910 addresses topics ranging from portable wood/metal ladders to the storage and handling of hazardous materials to automatic fire detection and sprinkler systems. One of the most critical OSHA regulations is 134, which focuses on respiratory protections; it is covered under 29CFR1910.134. This regulation

FIGURE 5.2 ◆ The fire service instructor needs to take many standards, rules, and regulations into account when conducting training evolutions.

establishes guidelines for fit-testing SCBA face pieces as well as operational requirements for interior fire fighting ("two-in/two-out"). The fire service instructor needs to be aware of the OSHA standards and comply with them. (See Figure 5.2.)

◆ NEGLIGENCE

Negligence is sometimes considered synonymous with malpractice. Although medical issues typically come to mind when thinking of malpractice, an instructor can also be considered negligent in performance or instruction. If someone in class dies or is injured, the instructor could be found negligent. (See Figure 5.3.) Four elements must be proven in order to be considered negligent:

1. *Duty to act:* The individual believed to be responsible had a legal obligation to act.
2. *Breach of duty:* The duty to act was breached by doing (committing) or not doing (omitting) a reasonable and prudent action.
3. *Injury:* An injury was sustained by the person who is suing.
4. *Cause (or causation):* A linkage exists between the injury that occurred and the breach of the duty to act.

If all four of these elements are proven, and the instructor is considered negligent for his/her actions, he/she can and will be held accountable. This could result in either criminal or civil proceedings.

FIGURE 5.3 ◆ The fire service instructor needs to pay attention to student safety. In this instance an instructor is checking a student ready to enter a burning building. *(Photo courtesy of James Clarke, Estero Fire Rescue)*

Criminal proceedings are typically instigated by a governmental agency and involve a penalty of either a fine and/or imprisonment—in some instances both. A person may be found not guilty or acquitted of criminal charges but still face civil charges. Civil action is when a party sues an individual personally, and any monetary penalties imposed are his/her responsibility. Many individuals purchase insurance policies to protect them financially against civil suits.

STANDARD FOR INSTRUCTION

A firefighter is expected to perform at a certain level. As an instructor the same holds true. The standard for fire fighting is what a "reasonable and prudent" individual who possesses similar training and experience would do in a similar situation. As an instructor, the standard will be what other fire service instructors would do under a similar situation. Some states define the standard of instruction by state law. The national standard for fire instructors is typically guided by NFPA standards. A number of states have formalized programs of instruction and processes for certification and review for instructors to ensure consistency and quality of instruction. For example, the state of Florida has three instructor levels; all three require a minimum of six years as a firefighter. In addition, Level I requires Fire Service Course Delivery, Level II requires Fire Service Course Delivery and Fire Service Course Design. Both Levels I and II require the individual to successfully pass a written test. The individual must also have an associate degree to be a Level II instructor. Level III requires Fire Service Course Delivery and Fire Service Course Design plus a bachelor's degree.

Some courses require individuals to take an instructor course to teach them the foundations of instruction. The courses have item-specific materials that require instructors to acquire knowledge or techniques essential. The instructor will need to inquire what the requirements are to instruct these specialized courses—and, of course, never teach courses that you are not qualified to instruct.

LIABILITY

A variety of areas of potential liability exist for instructors. The first is discrimination. An instructor is responsible for establishing a classroom format and environment that exemplifies professionalism. In Chapter 3 it was discussed that an instructor should understand there are many cultural, sociological, and economical factors that can play into students' conduct and attentiveness during the education and training process. The instructor should lead by example and take into account the varying backgrounds associated with cultural diversity. What may be funny to an individual or others may be offensive to one or more individuals in the classroom. Some instructors still want a "good ol' boy" classroom and think it is acceptable to tell off-color or racist jokes and make offensive gestures. This is completely unacceptable in this day and age.

An instructor needs to use consistent, fair practices for all the students. Listen first and then decide guilt or innocence using due process. Don't let personal bias enter into the decision. Instructors should have written documentation of every incident for their protection and the student's protection.

◆ HARASSMENT

Various behaviors constitute harassment, including embarrassment, demeaning, disgrace, humiliation, or intimidation of another. If a behavior acceptable to one is deemed by another as unacceptable, it is considered harassment. A policy and procedure needs to be established to govern the classroom environment. Everyone, including the students, needs to be aware of the policy.

Sample Classroom Modus Operandi

Policy

It is the policy of the instructors and students of this class to maintain an environment free from intimidation, coercion, or harassment, including sexual harassment, by any employee, student, or other individual while functioning in the capacity of an instructor or a student. Incidents of harassment by instructors, students, managers, vendors, or the public will not be tolerated and should be reported promptly to the training chief.

Procedure

Instructors are expected to conduct themselves in a businesslike manner at all times. Any behavior that is coercive, intimidating, harassing, or sexual in nature is inappropriate and prohibited. Any verbal, physical, or visual conduct that belittles or demeans an individual because of his or her race, religion, national origin, gender, age, disability, or similar characteristic or circumstance is prohibited.

Again, use consistent, fair practices for all the students. Bring in other instructors to assist, if necessary, but do not influence their objectivity with personal opinions.

SEXUAL HARASSMENT

Incidents of harassment may be subjective in nature. To assist instructors in understanding sexual harassment, the following is the federal government's definition in this policy:

Sexual Harassment Is. Unwelcome sexual advances, requests for sexual favors, and other physical, verbal, or visual conduct based on gender when:

1. Submission to the conduct is an explicit or implicit term or condition of employment,
2. Submission to or rejection of the conduct is used as the basis for an employment decision, or
3. The conduct has the purpose or effect of unreasonably interfering with an individual's work performance or of creating an intimidating, hostile, or offensive working environment.

Sexual harassment can include any of the following kinds of behavior:

Explicit sexual propositions
Sexual innuendo
Sexually suggestive comments
Sexually oriented teasing or kidding
Sexually oriented jokes
Obscene gestures or language
Unwanted hugs, kisses, touches
Retaliation for complaining about sexual harassment
Obscene or sexually suggestive pictures, publications, or drawings
Physical contact, such as patting, pinching, or touching

Instructors always need to be aware of how their actions may look to observers. What may seem innocent may be offensive to someone else. Avoid intimate situations or contact with students. Always avoid dating any student. After the class is completed, and you are no longer instructor/student, it is more acceptable. If things don't work out when dating a student, it may be an uncomfortable situation and even result in the student bringing a harassment issue against the instructor.

When counseling students, do so in a private setting, but take steps to prevent any unwanted situations. (See Figure 5.4.) A variety of options permit you to avoid these situations. The first would be to leave the classroom door open. Another option would be to have another instructor, preferably of the same gender as the student, sit in as a witness. Many classrooms have windows or windows in the door so that people can look in and see what is going on. If this is the case, the instructor should leave the blinds open for his/her protection and the student's.

Instructors are responsible for maintaining a classroom that is free of harassment, but all participants should help to ensure that harassment does not occur by conducting themselves appropriately and reporting any harassment they observe. If an individual has a complaint or allegation of harassment, he/she should immediately report the incident to the chief of training or the appropriate supervisor.

A complaint should be investigated promptly and the matter kept as confidential as possible. Retaliation of any kind against anyone who alleges harassment should be forbidden and any attempt to strike back should incur punitive action by the agency.

If after thoroughly investigating the agency determines that harassment has occurred, appropriate disciplinary action should be taken. Discipline may include guidance and counseling, written reprimand, suspension, revocation as an instructor with the agency, termination, or any other action deemed appropriate.

FIGURE 5.4 ◆ When counseling a student you should always be visible to avoid any potential accusations later. Having a witness in the room is always advisable.

GRIEVANCE PROCEDURES FOR STUDENTS

The fire service is accustomed to having a grievance procedure in place. The classroom is no different. If instructing for an academic institution, find out if there is a grievance procedure and become familiar with it. Students need to be provided written information on grievance procedures and due process in the student handbook or syllabus. If the instructor is not teaching for an academic institution and there is no procedure, a grievance procedure should be implemented, especially if granting certification or grades.

Students must be allowed to enact and go through the process without intimidation. Be sure to document incidents that have a potential to create a problem at the time of occurrence to protect against a grievance at a later time.

◆ ACADEMIC HONESTY ISSUES

It is important to establish the rules for the class up front so that students know what is expected of them. (See Figure 5.5.) The class an instructor is teaching will determine the depth of the rules. If operating a training center, the rules may be established in whole and be available to each student who attends a session. If instructing for a fire department, the department's SOPs/SOGs should cover any classroom setting. On the

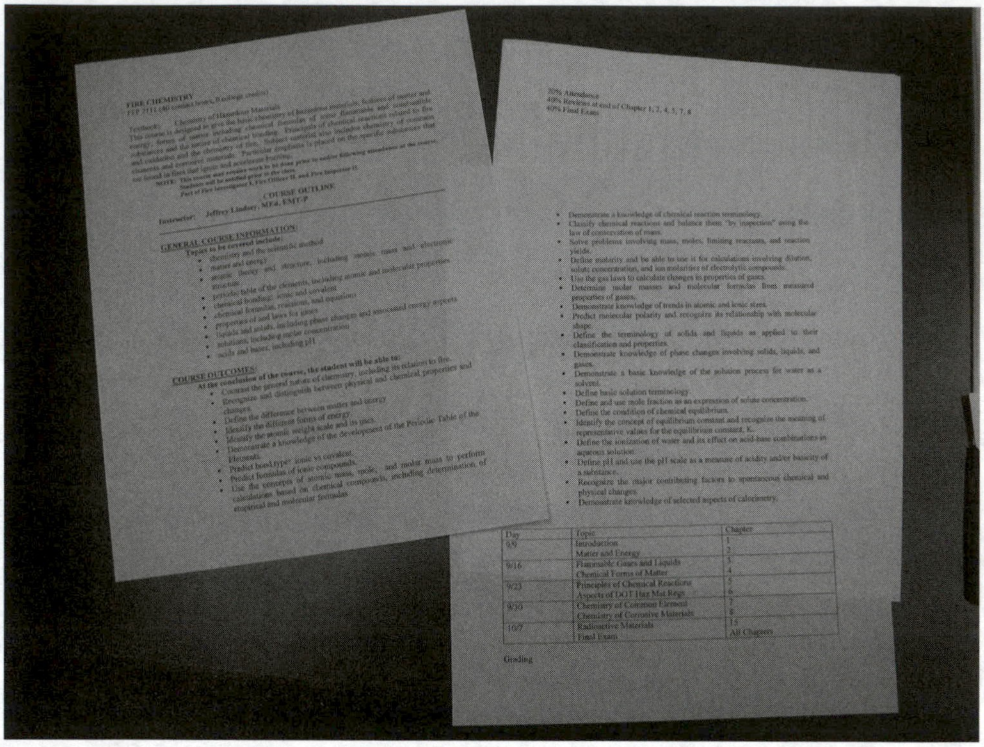

Figure 5.5 ◆ Your syllabus needs to be complete and detailed to establish the ground rules with the class.

other hand, if instructing a class in an academic setting, the rules should be stated in the syllabus. The written policies given to students should include:

1. *Academic Standards:* Academic standards should set forth the policy on grading. Make clear from the beginning what the requirements are for earning each letter grade or score. The policy should include how an individual can gain or lose points.
2. *Internet Usage:* If computers are available in the classroom, establish a policy regarding Internet usage. The Internet has opened many doors, both good and bad. An example of the latter are sites that sell term papers or essays to students. Counter to this are websites where a student's paper is submitted to see if it's plagiarized. One popular site is www.turnitin.com. If planning to use this service, address it at the start of the class, so students are aware of the process.
3. *Dishonesty Statement:* There needs to be a clearly written statement regarding what constitutes academic dishonesty. This includes cheating on exams, falsification of any work and experiences, logs, or other program documents, and attempts to reconstruct or obtain exam information.
4. *Affirmative Action/Equal Opportunity:* Prerequisites and entrance requirements must be fair and impartial. A provision must be in place for remedial or developmental education. Check with the institution or the fire department about the established policy to ensure compliance with the policy.
5. *Code of Conduct:* The fire service is governed by a code of conduct that can be emulated in the fire service classroom. The code of conduct should define the ethical and moral standards of the profession and should be applicable to instructors as well as students. It is important to remember that the student has rights and responsibilities, which need to be in writing.

In some settings, such as colleges and universities, students may actually have judicial powers. In instances where the instructor is employed by a fire department and under the guidance of a bargaining union, judicial powers may be allocated to the student. The instructor needs to be aware of the powers that are given to the student, and honor and respect the student's powers and govern himself/herself accordingly in the classroom environment.

Trainer Tales

In 1994 two firefighters narrowly escaped serious injury while conducting aerial training on the front ramp of their fire station. The firefighters were under the supervision of an experienced colleague, who was called to the phone. During his absence, the two firefighters, who were on the turntable of the ladder truck, decided to raise the ladder out of its bedded position. Due to an incline at the front ramp of the station, the turntable free-rotated—directly into high-voltage power lines that ran in front of the firehouse. (The two firefighters didn't know how to engage the manual turntable brake to prevent the ladder from rotating.) This energized the aerial ladder and burned a four-foot hole through the asphalt under one of the ladder's jack plates. Firefighters inside the station calmed the two firefighters on the ladder truck until the power company could de-energize the wires. The firefighters were unhurt—though the ladder truck sustained damage and had to be returned to the manufacturer for repair.

RISK MANAGEMENT CONSIDERATIONS

As discussed, firefighting is a high-risk occupation. Training sessions may also result in a high-risk environment. If teaching students who are not employees of a fire agency, encourage them to acquire student health insurance. If instructing for an educational institution, check with student services to see what insurance is available, since a number of programs cater to students. (See Figure 5.6.)

Students may be required to meet certain health requirements before enrolling in a course. For example, a number of fire academies require that fire cadets have a physical examination that certifies they are able to perform and conduct themselves in the practical evolutions to become a firefighter. Other institutions may require certain immunizations or other appropriate requirements.

Although instructor malpractice is not a common occurrence, instructor malpractice insurance covering errors and omissions should be considered. In most instances this coverage falls under the auspices of the employer's insurance; if an instructor needs to get coverage on his/her own, several insurance companies provide policies for this area. Ascertain what the coverage is for the policy purchasing. Some examples of what constitutes instructor malpractice are teaching misinformation or teaching a program that an instructor has developed with his/her own standards versus using a recognized standard.

CONFIDENTIALITY

Legislation has established and regulates criteria for the release of confidential information. Whereas an employee's training record is typically public record in the government setting, in an academic setting an instructor needs to pay attention to the laws of confidentiality. Occasionally a parent may want to view the records or discuss the progress of their child who is 18 years of age or older, but the records of

FIGURE 5.6 ◆ Fire fighting is a risky business. The proper insurances need to be acquired by the instructor, student, and the training center before beginning any training session. *(Photo courtesy of James Clarke, Estero Fire Rescue)*

students 18 or older are considered confidential, and parents don't have a legal right to see them. (See sidebar regarding the Family Educational Rights and Privacy Act of 1974.)

Confidentiality also is important to remember when discussing various topics or case studies in the classroom. In 1996 HIPAA (Health Insurance Portability and Accountability Act) created a new culture requiring compliance from the fire service. This is a very complex topic, and a variety of training programs cover it in depth. The bottom line when dealing with HIPAA is not to disclose any patient information if you are discussing any type of medical call in the classroom. This includes firefighters who are injured or killed.

Family Educational Rights and Privacy Act (FERPA)

The Family Educational Rights and Privacy Act (FERPA) (20 U.S.C. § 1232g; 34 CFR Part 99), also known as the Buckley Amendment, is a federal law that protects the privacy of student education records. The law applies to all schools that receive funds under an applicable program of the U.S. Department of Education.

FERPA gives parents certain rights with respect to their children's education records. These rights transfer to the student when he or she reaches the age of 18 or attends a school beyond high school level. Students to whom the rights have transferred are deemed "eligible students."

- Parents or eligible students have the right to inspect and review the student's education records maintained by the school. Schools are not required to provide copies of records unless, for reasons such as great distance, it is impossible for parents or eligible students to review the records. Schools may charge a fee for copies.
- Parents or eligible students have the right to request that a school correct records that they believe to be inaccurate or misleading. If the school decides not to amend the record, the parent or eligible student then has the right to a formal hearing. After the hearing, if the school still decides not to amend the record, the parent or eligible student has the right to place a statement within the record setting forth his or her view about the contested information.
- Generally, schools must have written permission from the parent or eligible student to release any information from a student's education record. However, FERPA allows schools to disclose those records without consent, to the following parties or under the following conditions (pursuant to 34 CFR § 99.31):
  1. School officials with legitimate educational interest;
  2. Other schools to which a student is transferring;
  3. Specified officials for audit or evaluation purposes;
  4. Appropriate parties in connection with financial aid to a student;
  5. Organizations conducting certain studies for or on behalf of the school;
  6. Accrediting organizations;
  7. To comply with a judicial order or lawfully issued subpoena;
  8. Appropriate officials in cases of health and safety emergencies; and
  9. State and local authorities, within a juvenile justice system, pursuant to specific state law.

Schools may disclose, without consent, "directory" information such as a student's name, address, telephone number, date and place of birth, honors and awards, and dates of attendance. However, schools must tell parents and eligible students about directory information and allow parents and eligible students a reasonable amount of time to request that the school not disclose directory information about them. Schools must notify parents and eligible students annually of their rights under FERPA. The actual means of notification (special letter, inclusion in a bulletin, student handbook, or newspaper article) is left to the discretion of each school.

For additional information or technical assistance, you may call (202) 260-3887 (voice). Individuals who use TDD may call the Federal Information Relay Service at 1-800-877-8339. Or you may contact them at the following address:

Family Policy Compliance Office
U.S. Department of Education
400 Maryland Avenue, SW
Washington, D.C. 20202-4605

PHOTO RELEASES

In most instances it's acceptable to take photos of individuals as long as they are in the public setting and the photo won't be used for commercial or marketing purposes, but it is essential to get a signed release when taking photos of students or others to use for marketing purposes. (See Figure 5.7 for a sample release form.)

Authorization and Consent to Use Photographs

The undersigned, intending to be legally bound, hereby authorizes Estero Fire Rescue, its agents and employees to use for any lawful purpose, including illustration, advertising, and publication in any manner, all photographs described below. The undersigned hereby irrevocably consents to the unrestricted use of such photographs and waives the right to inspect or approve any finished product or advertising copy that may be used in connection therewith.

DESCRIPTION OF PHOTOGRAPHS

Date of Photographs	General Description and Location

In witness whereof, the undersigned has executed this Authorization and Consent this __________ day of ____________________, 200__.

____
(print name)

____
(signature)

FIGURE 5.7 ◆ Sample form to use for authorizing the use of photographs

AMERICANS WITH DISABILITIES ACT (ADA)

Approximately 43 million Americans have a physical disability. For students with documented disabilities, certain reasonable accommodations must be made. These accommodations must be reasonable. In other words, if the accommodation represents something that would not be an expected element of job performance, but can be accommodated at a reasonable cost to implement, then it is recommended that the employer provide it.

The ADA is a federal antidiscriminatory statute designed to remove barriers that prevent qualified individuals with disabilities from enjoying the same opportunities available to persons without disabilities. An instructor can't discriminate against students or other instructors with disabilities. This prohibition covers all aspects of the employment process for instructors including:

- Application and promotion
- Testing and medical exams
- Hiring and layoff/recall
- Assignments and termination
- Evaluation and comprehension
- Disciplinary actions and leave
- Training and benefits

Employers are responsible for and required to make reasonable accommodations to qualified applicants or employees with disabilities. Some examples of reasonable accommodations include:

- Job restructuring
- Modifying work schedules
- Reassignment to another position
- Acquiring or modifying equipment or devices
- Adjusting or modifying examinations, training materials, or policies
- Providing qualified readers or interpreters
- Making existing facilities used by employees readily accessible to and usable by individuals with disabilities

Keep in mind that an instructor is not required nor expected to lower quality or quantity of standards to make an accommodation. Nor is the instructor required to provide personal items, such as glasses or hearing aids, as accommodations.

The Individuals with Disabilities Education Act is very clear as it regards learning disabilities, and who should be included in the transition services available. This act has hopefully identified those children with a learning disability and directed them toward the proper path for their learning years. In all likelihood, the individuals entering into the fire service profession will have recognized their learning disability.

Section 504 of the Rehabilitation Act of 1973 and the Americans with Disabilities Act of 1990 (ADA) protect individuals with learning disabilities from discrimination. Students who have documented their learning disabilities are entitled to accommodations to support their educational success. A student is responsible for making his/her learning disability known and for requesting adjustments to receive Section 504 accommodations. The types of accommodations include:

- Extended time on tests
- Note takers
- Assistance: use of technology devices (tape recorders or laptop computers)
- Modified assignments
- Alternative assessments and test formats

ACCOMMODATION EXAMPLE 1

The student can't read and has asked for an accommodation to have the test read to him. The instructor will of course take this matter to his/her administration (and perhaps their lawyers and the state fire marshal's office or attorney general's office) to solve, but most likely the instructor won't have to provide this accommodation because reading ability is a requirement for the profession.

FIGURE 5.8 ◆ On December 9, 1999, Brian Lutz was involved in an accident. He was driving an ambulance on an emergency response. His unit rolled over and Brian was partially ejected from the vehicle. He was rendered unconscious and flown to the regional trauma center. A few days later he was faced with a decision of whether to chance many operations and long rehabilitation or have his right leg amputated. He elected to have his right leg amputated. Brian went through rehab and performed the physical agility test to return to work. Brian was accommodated with a left-foot accelerator for his unit, and he was back on the street driving ambulances a little over 8 months from the time of his accident. Brian is shown being interviewed at a conference by Rick Patrick of VFIS.

ACCOMMODATION EXAMPLE 2

The student has documentation diagnosing dyslexia from a physician. She is able to process information if given a little longer to take written tests. Again the instructor consults with his/her administration, and the administration rules that it is acceptable to add some additional time to the written test because there is no standard in the fire service requiring how fast a person must be able to read.

As a fire service instructor, do not assume just because someone has a disability he/she cannot do the job. There are laws preventing discriminating against someone with a disability. (See Figure 5.8.)

◆ COPYRIGHT ISSUES

Instructor can no doubt recall the many outside resources that were used in the classroom through the years of their own education. Outside resources have been used by instructors to enhance students' learning experience for as many years as there have been resources. These can range from a newspaper article illustrating an incident, to a book to highlight a skill, to a photograph of a fire scene. These and other resources are copyrighted materials.

The original copyright law was created in 1790 and was referred to as the Copyright Act of 1790. Recently Congress enacted the Sonny Bono Copyright Term Extension Act of 1998. This law provides for the proper use of copyrighted materials, especially in educational settings where such materials fall under what is referred to as "fair use." A number of factors are used in the fair-use test, and the standards for usage of materials in education are more general than they are for public use. (It should be noted that the application of the federal law may vary state by state and locale by locale.)

Fair Use

In determining whether copying of someone else's work is permitted by the doctrine of fair use, Section 107 of the Copyright Law lists the four factors that courts are to consider in defining "fair use" (17 USC section 107). These are

1. *Purpose and Character:* What is the purpose and character of the copying? Is the intended use of a commercial nature, or is it for nonprofit educational purposes? For example, copying for nonprofit educational uses will more likely qualify as fair use.
2. *Nature of Work:* What is the nature of the work being copied? For example, copying from works that are primarily factual is tolerated more than copying from more creative works.
3. *Amount and Quantity Copied:* What is the amount and substantiality of the portion that is copied in relation to the copyrighted work as a whole? For example, the more that is copied or the more significant the portion that is copied (regardless of the quantity), the less likely fair use will apply.
4. *Effect on Market:* What is the effect of the copying on the potential market for or value of the copyrighted work? For example, if the copying has an adverse impact on the market for the original work, it will not constitute fair use.

Exemptions

Besides fair use, schools also benefit from some highly specific exemptions from copyright liability. For example, Section 110 of the Copyright Act provides several specific limitations on the exclusive rights that copyright owners enjoy for their works. These limitations permit schools and educational institutions to display and perform copyrighted works under very specific circumstances.

In face-to-face teaching activities, Section 110 (1) permits instructors or pupils at nonprofit educational institutions to perform or display copyrighted works so long as the work is performed or displayed in a classroom or similar place devoted to instruction. The educational exemption for using copyrighted works transmitted by any device or process is much more limited.

A copyright grants the holder the sole right to reproduce or give permission to others to reproduce the copyrighted works. The copyright holder is defined as the person who owns the exclusive rights to a work. The protection is limited to original works, whether or not they have been published.

For works created prior to January 1, 1978, copyright protection lasts 95 years from the date of first publication or 100 years from the date of creation of the work, contingent on which date allows the copyright to expire first for works of corporate authorship. For works created after 1978, protection begins at the creation of the work and lasts 75 years after the death of the author (17 USC Section 104A).

The following guidelines indicate what copying is allowed when the instructor decides to use a work spontaneously for educational purposes (NACS and AAP, 1991) and also outline the length of works that may be copied:

- A complete article or story less than 2,500 words;
- 1,000 words or 10 percent (whichever is shorter) of a prose work that is excerpted;
- One illustration, chart, diagram, or picture per book or periodical issue; and
- A short poem of less than 250 words or an excerpt of a longer poem of not more than 250 words.

Digital Millennium Copyright Act (1998)

The Digital Millennium Copyright Act of 1998 was enacted to cover the copyright issues regarding digital transmission of information on October 28, 1998. (See Figure 5.9.) It is important to note that the law is still being defined and tested in court. However, the highlights of the act are

- It is a crime to circumvent antipiracy measures built into most commercial software.
- The manufacture, sale, or distribution of code-cracking devices used to illegally copy software is outlawed.

FIGURE 5.9 ◆ The Internet has a wealth of information. Be sure to consider copyright issues when using items from the Web.

- The cracking of copyright protection devices *is* permitted to conduct encryption research, assess product interoperability, and test computer security systems.
- Under certain circumstances nonprofit libraries, archives, and educational institutions are exempt from anticircumvention provisions.
- In general, Internet service providers (ISPs) are limited from copyright infringement liability for simply transmitting information over the Internet.
- ISPs are expected, however, to remove material from users' websites that appears to constitute copyright infringement.
- When nonprofit institutions of higher education serve as online service providers—and under certain circumstances—they have limited liability for copyright infringement by faculty members or graduate students.
- "Webcasters" are required to pay licensing fees to record companies.
- After consultation with relevant parties, the Register of Copyrights is required to submit to Congress recommendations regarding how to promote distance education through digital technologies while "maintaining an appropriate balance between the rights of copyright owners and the needs of users."
- The Act states explicitly: "Nothing in this section shall affect rights, remedies, limitations, or defenses to copyright infringement, including fair use. . . ."

(*Source:* The Digital Millennium Copyright Act—Overview, *http://www.gseis.ucla.edu/iclp/dmcal.htm*)

Obtaining Copyright Permission

Copyright permission must be obtained from the copyright holder of the work in which the instructor is interested. Copyright notice is optional for works published on or after March 1, 1989, however, so tracking down the copyright holder may be difficult (NACS and AAP, 1991).

When requesting copyright permission, include all the following information (NACS and AAP, 1991):

- Full name(s) of the author, editor, and/or translator
- Title, edition, and volume number of the work
- Copyright date of the work
- ISBN for books or ISSN for magazines
- Exact pages, figures, and illustrations you wish to use
- The number of copies to be made
- Whether the material will be used alone or in combination with other works
- Name of the college or university
- Date when the material will be used
- Instructor's full name, address, and telephone number

COPYRIGHT AND INTELLECTUAL PROPERTY ISSUES

It is sometimes assumed that if a document does not have the copyright symbol, it is not copyrighted. A good rule of thumb is that any document you did not author is typically owned by somebody else. Depending on the work and the source, the person who owns it or develops the material deserves credit, and in some instances even compensation, for the work if you elect to use it. "Public domain" is any work that is exempt from copyright laws because of the age of the document or the information is considered to be known by most individuals.

In some instances you may not know who authored a document or be able to find who to get the permission from to use the material. At least make a good-faith effort to obtain permission to use any document that is not yours.

Conclusion

A basic knowledge of copyright law is essential for any fire service instructor. (Instructing in a distance-education setting adds other concerns due to the nature of the educational environment.) This issue concerns the course developer, instructor, the students, the administration, and the institution with which they are all involved—and it is not an issue that can be taken lightly.

Note: Copyright law and intellectual property rights are extremely complex issues. Questions regarding a specific circumstance should be directed to legal counsel.

SOURCES FOR ADDITIONAL INFORMATION ON FIRE LAWS

There are many resources available to you as a fire service instructor. (See Figure 5.10.)

State fire marshal's office
Federal government agencies dealing with regulation and oversight
National organizations
International Fire Chief's Association
National Fire Protection Association
United States Fire Administration
Trade journals for fire
Books on fire law
Internet

FIGURE 5.10 ◆ Examples of resources

SUMMARY

This chapter has provided an overview of legal issues. The instructor should consult the legal department for in-depth information and consultation on matters involving any legal issues pertaining to the course or program.

Review Questions

1. Describe at least one standard that affects the fire service instructor.
2. Define the four points of negligence.
3. Describe a situation that would result in a fire service instructor being charged with negligence.
4. Describe what constitutes sexual harassment.
5. Describe academic dishonesty.
6. Describe the Buckley Amendment.
7. Describe some reasonable accommodations for students with a disability.
8. Describe fair use as it relates to copyright.
9. List at least four other resources to find related fire-related law and instruction.

References

National Highway Safety Transportation Administration. August, 2002. *National guidelines for educating EMS instructors*. Washington DC.

17 United States Code (USC). U.S. Copyright Act, As Amended. Washington, DC.

National Fire Protection Association. How the code process works, accessed November, 17, 2004, *http://www.nfpa.org/categorylist.asp?CategoryID=162&URLCODES%20and%*.

Bibliography

Copyright and Distance Education. University of Idaho. Retrieved on October 12, 2003, from *http://www.uidaho.edu/evo/dist13.html*.

Distance education: strategies and tools. educational technology publications. Englewood Cliffs, NJ: Prentice Hall.

Dalziel, C. 1995. "Copyright and you: Fair use guidelines for distance education." *Techtrends,* October, 6–8.

Flight, M. 1998. *Law, Liability, and Ethics* (3d ed.). New York: Delmar Publishers.

Mason, A. 1996. *Copyright and trademark law*.

National Association of College Stores (NACS), Inc. and Association of American Publishers (AAP). 1991. *Questions and answers on copyright for the campus community*. Oberlin, OH: National Association of College Stores, Inc.

House Report (HR) No. 1476. 1976. 94th Congress, 2nd Session.

CHAPTER 6

Conducting Practical Training Exercises

Terminal Objective

The participant will be able to conduct a live fire training session as defined by NFPA 1403.

Enabling Objectives

- Identify the NFPA standards associated with conducting live fire training.
- Identify the agencies involved in conducting a live fire training session.
- Describe the process of conducting a live fire burn on an acquired structure.
- Describe the various components you need to establish at a live fire burn.
- Define the safety procedures to implement a live burn.
- Define and describe the key elements of post-incident analysis.
- Describe the documentation you need to conduct a live fire burn at an acquired structure.
- List and describe the various types of interior and exterior simulator props.

JPR NFPA 1041—Instructor I

4-2.2 Assemble course materials given a specific topic so that the lesson plan, all materials, resources, and equipment needed to deliver the lesson are obtained.

4-2.3 Prepare training records and report forms, given policies, and procedures and forms so that required reports are accurately completed and submitted in accordance with the procedures.

4-3.2 Review instructional materials, given the materials for a specific topic, target audience, and learning environment, so that elements of the lesson plan, learning environment, and resources that need adaptation are identified.

4-4.2 Organize the classroom, laboratory, or outdoor learning environment, given a facility and an assignment, so that lighting, distractions, climate control, or weather, noise control, seating, audiovisual equipment, teaching aids, and safety are considered.

JPR NFPA 1041—Instructor II

5-2.4 Acquire training resources, given an identified need, so that the resources are obtained within established timelines, budget constraints, and according to agency policy

5-2.5 Coordinate training record keeping, given training forms, department policy, and training activity, so that all agency and legal requirements are met.

5-4.3 Supervise other instructors and students during high hazard training, given a training scenario with increased hazard exposure, so that applicable safety standards and practices are followed and instructional goals are met.

JPR NFPA 1041—Instructor III

6-2.4 Select instructional staff, given personnel qualifications, instructional requirements, and agency policies and procedures, so that staff selection meets agency policies and achievement of agency and instructional goals.

◆ INTRODUCTION

It is becoming more and more difficult to conduct live burns. (See Figure 6.1.) A number of issues may preclude or deter the use of structures to conduct live fire training, including the number of injuries and fatalities as a result of live burns, and environmental factors. Hence, using fire simulators is becoming more prevalent in fire training.

Trainer Tales

Deviation from an established training procedure resulted in a near tragedy during a fire training exercise. In this incident, fire academy staff applied drywall to two rooms of a structure that was acquired to conduct interior fire-attack training. Per local environmental and departmental health regulations, the asbestos had been stripped from all walls and other enclosures.

The interior staff, consisting of the safety officer and the igniter, started a fire outside the dry-walled area for an "orientation" fire to practice limited interior attack. The fire rapidly progressed to the attic and roof, and the structure quickly became heavily involved. The interior crew had a properly connected safety-hose line, but it was in the dry-walled area where the fire was now raging. The crew was cut off from the safety line as well as their primary means of escape, and they had to retreat out a bathroom window. This close call, caused by deviation from an established procedure, resulted in only minor burns—but it had the potential for disastrous results.

Figure 6.1 ◆ Live fire burns can provide a great training opportunity for new as well as experienced firefighters. There are a number of precautions you need to consider when conducting live fire burns.

◆ LEGAL ISSUES INVOLVING LIVE BURNS

A number of NFPA standards can be referenced as regards any type of live fire burn. They include: NFPA 1403, Standards on Live Fire Training Evolutions; NFPA 1001, Standard for Firefighter Professional Qualifications; NFPA 58, Liquefied Petroleum Gas Code; NFPA 59, Utility LP Plant Code; NFPA 1981, Standard on Open-Circuit Self-Contained Breathing Apparatus for the Fire Service; and NFPA 1982, Standard on Personal Alert Safety Systems. Other standards may be applicable depending on the training the instructor is conducting. The NFPA standards are typically considered to be the national standards regulating and governing the fire service. State and local laws also govern live-fire burns. Instructors will need to check with the jurisdictions in their area to determine what other laws affect live fire training.

The purpose of NFPA 1403 is to establish requirements to provide a safe environment for the participants and the public while conducting live fire training. (It is important to note that this standard does not cover aircraft, marine structures and vessels, and ground cover and wildland fire training.) Chapter 1 of the document discusses the scope and purpose of the standard, which defines the minimum requirements for live fire training. Chapter 2 identifies the reference publications of the standard, the two main references being NFPA and NIOSH. Chapter 3 includes the terms used in this standard and their definitions.

In Chapter 4 the standard discusses the requirements for both acquiring structures and preparing for the burn. It focuses on the participants, structure/facilities, and the instructors who conduct live fire training. An instructor needs to refer to NFPA 1001 regarding preparing participants. Keep in mind that participants need to wear PPE, SCBA, and a PASS device during live fire training. It is important to note that there needs to be a 5:1 ratio of students to instructors for live fire training.

Chapter 5 gets into gas-fired training center buildings. It sets forth the requirements for participants, structures/facilities, fuel materials, safety officers, and instructors. Discussed in this chapter are regularly scheduled inspections that are conducted at least annually. A dedicated safety officer, to prevent and assist in eliminating unsafe acts, is required during burns. This individual also has the authority to cease the training immediately if it is deemed to be unsafe at any time. The chapter also details the construction of a gas-fired burn building and the requirements associated to maintain it. (See Figure 6.2.)

Chapter 6 establishes the standards for nongas-fired training center buildings. It sets forth the requirements for participants, structures/facilities, fuel materials, safety officers, and instructors. It, too, cites the requirements for construction and maintenance of the building.

Chapter 7 deals with exterior props. The chapter continues the format of setting forth the requirements for participants, structures/facilities, fuel materials, safety officers, and instructors.

FIGURE 6.2 ◆ Gas-simulated fire props have a number of controls and safety switches. Instructors need to be familiar with running a gas-simulated fire evolution.

Chapter 8 covers external Class B fires. In addition to describing the requirements for participants, structures/facilities, fuel materials, safety officers, and instructors, this chapter also gets into the requirements for using flammable liquids and hydrocarbons that are "controllable."

Chapter 9 discusses the requirements for reports and records that are part of live-control training burns. The elements of record-keeping include but may not be limited to: activities, instructors (with assignments), participants, unusual conditions, injuries, changes in structure, final report, and post-incident analysis (PIA).

At the end of the standard is a section on commonalities that includes the following topics:

- Preparation of site
- Predrill plan (including emergency plan)
- Safety officer(s)
- Established water supply
- Safety hand-line—dedicated separate water source
- In-charge instructor—"Burn Master"
- Fuel
- EMS on site
- Ignition officer
- Rehab
- PIA

LOCAL AND STATE AGENCIES

A number of governmental agencies may be involved in live burn training. The instructors need to check the rules and regulations within their state and municipality to determine what other legalities they will need to follow. Some of the agencies that may be involved include:

- Building department
- Environmental Protection Agency
- Utilities
- Department of Agriculture
- Code Enforcement
- Department of Environmental Protection
- State or local police

Safety and legality issues are a critical element in the role of live fire burning. Make sure to check with all agencies to be sure of compliance with the laws, rules, regulations, standards, and ordinances set forth in the jurisdiction the burning will occur.

◆ DESIGNING AND PLANNING A LIVE BURN

The first step in conducting a live burn is to design and plan the event. As an instructor everything you do should consist of careful design and proper planning. These elements are especially crucial when doing a live fire burn.

DOCUMENTATION

If the training agency is fortunate enough to have a legal department, the instructor should seek assistance in locating and completing the documentation necessary to conduct a live burn. Everything done in the fire service requires some type of written documentation and conducting a live burn is no exception. The minimum required legal documents consist of permission from the owner(s), and a copy of the documented clear title for the structure to burn. Also ensure the owner has cancelled any insurance on the structure. Plus a burn permit must be secured from the local government where the structure stands. Figure 6.3(A)–(K) shows samples of live fire burn documents needed to conduct a live fire burn.

ORGANIZATION ASSIGNMENT LIST	
1. Incident Name Block Lane Training	
2. Date June 18, 2003	3. Time 0800
4. Operational Period 0800-1700	
Position	Name
5. Incident Commander and Staff	
Incident Commander	Steve Harris
Deputy	Dwyer
Safety Officer	Lindsey / Clemens
Information Officer	Cato
Liaison Offier	Linsey
6. Agency Representative	
Agency	Name
San Carlos	PHil Blanc
Iona McGegor	John Spicuzza
Bonita	T.B.A.
LCEMS	Medic 21
7. Planning Section	
Chief	Ed Dwyer
Deputy	
Resources Unit	
Situation Unit	
Documentation Unit	
Demobilization Unit	
Technical Specialists	
Human Resources	
Training	
8. Logistics Section	
Chief	
Deputy	
Supply Unit	
Facilities Unit	
Ground Support Unit	
Communications Unit	
Medical Unit	LCEMS Medic 21/ Clemens
Security Unit	
Food Unit	

9. Operations Section		
Chief	Battalion A, B, C.	
Deputy		
a. Branch I - Division/Groups		
Branch Director	Battalion Chief Nisbet	
Deputy		
Division/Group	Althouse	A-1
Division/Group	Poole	A-3
Division/Group	Wahlig	A-4
Division/Group		
Division/Group		
b. Branch II - Division/Groups		
Branch Director	Battalion Chief Vanderbrook	
Deputy		
Division/Group	Horton	B-1
Division/Group	Reisen	B-3
Division/Group	Coppell	B-4
Division/Group		
Division/Group		
c. Branch III - Division/Groups		
Branch Director	Battalion Chief Krajic	
Deputy		
Division/Group	Primmer	C-1
Division/Group	Stevens	C-3
Division/Group	Davis	C-4
Division/Group		
Division/Group		
d. Air Operations Branch		
Air Operations Branch Director	N/A	
Air Attack Supervisor		
Air Support Supervisor		
Helicopter Coordinator		
Air Tanker Coordinator		
10. Finance Section		
Chief		
Deputy		
Time Unit		
Procurement Unit		
Compensation/Claims Unit		
Cost Unit		
Prepared by (Resource Unit Leader) Battalion Chief Nisbet		

ICS 203 NFES 1327

FIGURE 6.3(A)

Estero FIRE RESCUE
19850 Breckenridge Drive, Suite A
Estero, Florida 33928
'hone: (941) 947-FIRE (3473) Fax: (941) 947-9538

web site: http://www.esterofire.org

889

ESTABLISHED 1984

BURN SITE INSPECTION
Acting as Agent for the Florida Division of Forestry

Burn Site

Owners Name: ________

Address: ______ Black Lane

Telephone #: 947- 3334

Forestry Control # ________

Section ________
Township ________
Range ________

Agricultural ( ) **Prescribed Burn (X)**

Open Pit ( ) **Air Curtain ( )**

Items to be burned: House

____

____

____

Dates and Hours: 6- -03

Inspection Denial (Reasons) ________

____

____

Receipt for Fee Received – Amount $ 0

Date Received: ________

Dailey Control Numbers: ________

____

____

Applicant

Name: ESTERO FIRE DEPT.

Address: 19850 Breckenridge Dr.

Telephone #: 947-3473

Emergency # ________

Applicant Control # ________

X ________
Signature

6-16-03
Date

Conditions of Inspection

Full compliance with Estero Fire Rescue and Florida Forest Fire Laws and Open Burning Regulations. Failure to comply with these provisions may result in the revocation of any authorized burn permit.

It is mutually understood and agreed that this inspection was conducted to determine compliance with the terms and conditions as set forth under Estero Fire Rescueand Florida Fire Laws and Open Burning Regulations and, furtherthat the applicant, by his/her signature hereon, does acknowledge and agree that the Estero Fire Rescue does not, by reason of granting site approval upon the terms and conditions herein stated, assume any responsibility or liability to any person or propertywhich might arise out of the burning permitted hereunder, but to the contrary, any and all liability and responsibility for any injury to any person, persons, firms or corporations, or to a property, which may arise out of this burning, shall be strictly that of the applicant.

X ________ 6/16/03
Applicant Signature and Date

X D. Burke 6/16/03
Estero Fire Rescue Signature and Date

Telephone Numbers:
Estero Fire Rescue Admin Ofc: 947-3473
Estero Fire Rescue Station 1: 947-5338
Division of Forestry: 694-5579

FIGURE 6.3(B)

Additionally, liability insurance specific to conducting a live burn is needed. A number of insurance agencies include this as part of an organization's insurance coverage; however, it is important to be 100 percent sure there is coverage prior to commencing the exercise. If other agencies are participating in the live burn, appropriate

DIVISION ASSIGNMENT LIST	1. Branch Battalion A	2. Division/Group A1, A3, A4
3. Incident Name Block Ln. Training	4. Operational Period Date: 06/18/2003 Time: 0800-1600	

5. Operations Personnel			
Operations Chief	Battalion Chiefs	Division/Group Supervisor	
Branch Director		Air Attack Supervisor No.	

Strike Team/Task Force/ Resource Designator	Leader	Number Persons	Trans. Needed	Drop Off PT./Time	Pick Up PT./Time
A1	Althouse	3			
A3	Poole	3			
A4	Wahlig	4			

6. Resources Assigned this Period

7. Control Operations

1. Maintain Crew accountability and Safety at all Times.
2. Provide instruction in hose advancement and Placement.
3. Experience changes in fire invironment by the progression of fire within a structure.
4. Practice water application in offinsive and defensive operation
5. Practice the operation of fire pumps while involved in firefighting operations

8. Special Instructions

1. Company officers will maintain their crews at all times.
2. Battalion Chiefs will rotate with crews to insure safety, and instruction consistency.
3. A back-up line will be staffed and ready to go at all times that crews are in the structure.
4. Personnel will be rotated to insure proper rehab, and instruction.

9. Division/Group Communication Summary

Function	Frequency	System	Channel	Function	Frequency	System	Channel
Command	Tac.7	King NIFC	Disp. S.	Logistics		King NIFC	
Tactical Div/Group	Tac. 7	King NIFC		Air to Ground		King NIFC	

Prepared by (Resource Unit Leader)	Approved by (Planning Section Chief)	Date	Time

FIGURE 6.3(C)

documentation will need to be secured from each agency. It is also imperative that the credentials of those conducting the burn are well documented, including their fire instructor certificates.

PREBURN PLANNING

When beginning the planning stage of the live burn, start by either obtaining or drawing a site plan. On the site plan be sure to include all the details. Some details to include would be the buildings, driveways, hydrants, utilities, and trees. It is important to also note all the exposures, not just for the fire but for smoke travel. If it's a windy day,

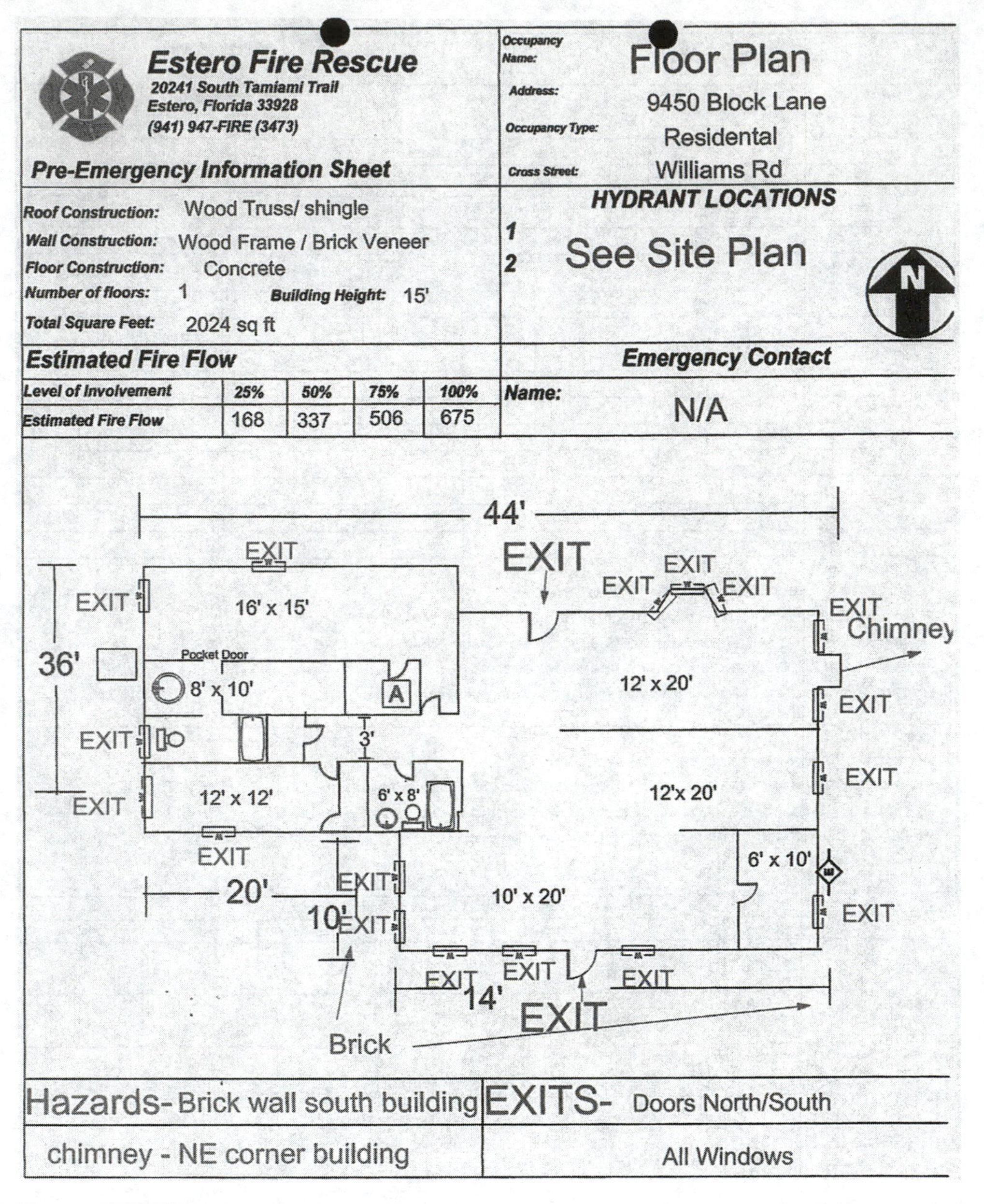

Estero Fire Rescue
20241 South Tamiami Trail
Estero, Florida 33928
(941) 947-FIRE (3473)

Pre-Emergency Information Sheet

Occupancy Name: Floor Plan
Address: 9450 Block Lane
Occupancy Type: Residental
Cross Street: Williams Rd

Roof Construction: Wood Truss/ shingle
Wall Construction: Wood Frame / Brick Veneer
Floor Construction: Concrete
Number of floors: 1 Building Height: 15'
Total Square Feet: 2024 sq ft

HYDRANT LOCATIONS
1
2 See Site Plan

Estimated Fire Flow

Level of Involvement	25%	50%	75%	100%
Estimated Fire Flow	168	337	506	675

Emergency Contact
Name: N/A

Hazards- Brick wall south building
chimney - NE corner building

EXITS- Doors North/South
All Windows

FIGURE 6.3(D)

what is in the path of the smoke if the wind is blowing in that direction? The site plan will be used for locating the command post, staging area, and other designated parts of the training evolution. Planning the burn is a good exercise not only on fire behavior and attack but also on the skills and knowledge to properly and effectively plan a structure in an event of an incident. Part of the planning process should include placement of vehicles. It is also important to note the surrounding area and speak to any

Estero Fire Rescue
20241 South Tamiami Trail
Estero, Florida 33928
(941) 947-FIRE (3473)

Pre-Emergency Information Sheet

Occupancy Name: Pre-Burn Plan
Address: 9450 Block Lane
Occupancy Type: Residence
Cross Street: Williams Rd

Roof Construction:
Wall Construction:
Floor Construction:
Number of floors: Building Height:
Total Square Feet:

HYDRANT LOCATIONS
1 across st from on River Ranch
2 NE corner of church

N

Estimated Fire Flow

Level of Involvement	25%	50%	75%	100%
Estimated Fire Flow				

Emergency Contact
Name:

Church
E212
5"
1 3/4"
1 3/4"
3"
Rehab
3"
E230
Blitz fire
T730
1 3/4"
E211
3"
IC
2 1/2"
House
House
Xtra 100' section 5"
10 sq acres

Sprinkler: FDC Location
X

Lock Box: Location
X

Standpipe: FDC Location
X

FIGURE 6.3(E)

close neighbors. They need to be apprised about the number of vehicles that will be in the area, the smoke, and noise from the training.

The next part of the process would be to obtain or draw the building plan that should include a detailed floor plan for each floor. Include all pertinent details, such as doors, windows, and stairways; additionally, each of the rooms needs to be labeled. As part of the safety efforts in planning, location of the fires that will be set should be planned, along with the emergency escape routes for each of these fires.

INCIDENT BRIEFING	1. Incident Name Block Lane Training	2. Date June 18,2003	3. Time 0800

4. Map Sketch

SEE ATTACHED

5. Current Organization

Incident Commander: S. HARRIS

Safety Officer: LINDSEY
Liaison Officer or Agency Rep: SPALLIZA, BLANCI T.B.A
Information Officer: CATO

Planning	Operations	Logistics	Finance
DWYER	Battalion A, B, C		

Div. A-SHIFT	Div. B-SHIFT	Div. C-SHIFT	Div. (D)	Air
A-1 - Althouse	B1 - Horton	C1 - Primmer	Bonita	Air Operations ______
A-3 - Poole	B3 - Rosen	C3 - Stevens	Iona	Air Support ______
A-4 - Wahlig	B4 - Coppell	C4 - Davis	SanCarlose	Air Attack ______
				Air Tanker Coord ______
				Helicopter Coord ______

Page 1 of	6. Prepared by (Name and Position) Battalion Chief Nisbet

NFES 1325

FIGURE 6.3(F)

During planning, the number and type of apparatus that will be needed should be determined. If the apparatus and personnel will be limited, the types of fires and number of evolutions will need to be limited to coordinate with this number. In the same respect, if there is a small building to burn, having more personnel at the event than the amount of training time can accommodate is not appropriate.

INCIDENT OBJECTIVES

1. Incident Name	2. D[illegible]	3. Time
Block Lane Training	June 18, 2003	0800

4. Operational Period
0800-1600

5. General Control Objectives for the Incident (include alternatives)

1. Firefighter Safety is of the Highest Priority.
2. Company level offensive and defensive firefighting strategies and tactics.
3. Implemntation of the Incident Command System and Accountability System.
4. Deployment of appropriate size attack lines.
5. Proper supply and pressures for attack lines to maintain optimal application rates.
6. Experience live fire conditions while increasing confidence in operation.
7. Provide a consistency between all personnel on fire ground command and control.
8. Practice the deployment of differn sized attack lines
9. Observe fire growth and fire characteristics

6. Weather Forecast for Period

7. General Safety Message

1. All personnel are required to check in upon arrival to the training site.
2. All personnel are to remain in a stageing location until assigned by an officer.
3. Company officers will maintain their crews intact while performing in the Hot Zone, a B/C will accompany hoselines.
4. Any and all personnel will report any unsafe condition, or change in conditions that might provide a hazard.
5. All personnel are to hydrate themselves as much as possible.
6. A Personal Accountability Report (PAR), will be conducted after each crew leaves the fire building or every 20 minutes.
7. All Personnel are required to rehab, and check out before leaving for the day.

8. Attachments (mark if attached)

☒ Organization List - ICS 203	☒ Medical Plan - ICS 206	☒ (Other) Letter
☒ Div. Assignment Lists - ICS 204	☒ Incident Map	☐
☒ Communications Plan - ICS 205	☐ Traffic Plan	☐

9. Prepared by (Planning Section Chief)	10. Approved by (Incident Commander)
B/C Nisbet	

FIGURE 6.3(G)

Water supply is an important element, one that needs to synchronize with the required fire flow. (See Figure 6.4.) After the fire flow has been determined, the water supply will need to be checked to ensure that you have an ample amount. Assess for hydrant supply, nearby water sources for drafting, and water shuttles. The amount of reserve water supply should be 50 percent more than the required water supply. For further information on the standards pertaining to water supply, refer to NFPA 1142, Standard on Water Supplies for Suburban and Rural Firefighting.

Weather is another factor to consider when in the plan. Consider the typical weather pattern for that time of year when the date for the burn is initially scheduled. If the date is in the middle of winter in the northern Midwest, consideration needs to

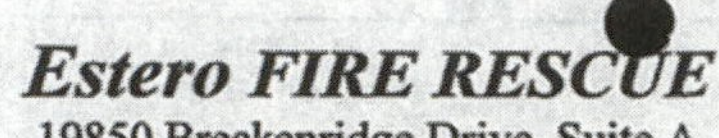

Estero FIRE RESCUE
19850 Breckenridge Drive, Suite A
Estero, Florida 33928

Phone: (239) 947-FIRE (3473) Fax: (239) 947-9538

web site: www.esterofire.org

June 10, 2003

To: Block Lane Residents
From: Battalion Chief Lawrence Nisbet
Regarding: Training at 9450 Block LN.

To whom it may concern,

Estero Fire Rescue will be conducting training on the property located at 9450 Block LN.. This training will involve several fire units and firefighting personnel, and will ultimately lead to the burning down of the structures located on the property. The training is scheduled to begin promptly at 8:00 a.m., and should be completed by 4:00 p.m.. During this training there will be hose deployed along Block LN., and there may be smoke in the area. Estero Fire Rescue appreciates your assistance on allowing us to have this training. Providing this training will enhance our firefighters readiness in handling future fire emergencies. If you have any questions concerning this training, please feel free to contact me at 390-1650.

Sincerely,

Lawrence Nisbet
Lawrence Nisbet
Battalion Chief

FIGURE 6.3(H)

be made for cold and inclement weather conditions. Likewise, if the date falls in the summer months and the evolution is being conducted in the Desert Southwest with temperatures soaring into the 100s, be prepared for heat exhaustion. It is recommended to plan the burn during a time when extreme weather conditions are not typical. Of course, it's difficult to plan for the weather on a particular date so far in advance; however, as the day gets closer, the weather should be monitored, and a second date should be chosen in case the original date needs to be canceled. The weather should also be monitored on the day of the burn. Special consideration needs to be noted for wind direction and any severe weather that may affect operations. Personnel should not be subjected to severe weather in a training evolution.

Other areas to consider when planning for the live fire burn or any practical evolution are various incident command functions. These include participant staging, operations, communications, and rehab. Designate a staging area so that participants can

MEDICAL PLAN	1. Incident Name	2. Date Prepared	3. Time Prepared	4. Operational Period
	Block Lane Training	June 18, 2003	0800	0800-1600

5. Incident Medical Aid Station

Medical Aid Stations	Location	Paramedics Yes	Paramedics No
Located at the Stageing Area		X	

6. Transportation

A. Ambulance Services

Name	Address	Phone	Paramedics Yes	Paramedics No
LCEMS Medic 21	EFR Station 3		X	

B. Incident Ambulances

Name	Location	Paramedics Yes	Paramedics No
LCEMS Medic 21	EFR Station3	X	

7. Hospitals

Name	Address	Travel Time Air	Travel Time Ground	Phone	Helipad Yes	Helipad No	Burn Center Yes	Burn Center No
Gulf Coast			10		X			X
Lee Memorial		7	20		X			X

8. Medical Emergency Procedures

All Medical Emergencies will be handled by EFR Station 3 personnel. LCEMS Medic 21 will be on-site to provide assistance and transport.

Prepared by (Medical Unit Leader)	10. Reviewed by (Safety Officer)

ICS 206

FIGURE 6.3(I)

INCIDENT RADIO COMMUNICATIONS PLAN		1. Incident Name Block Lane Training		3. Operational Period Date/Time June 18,	
4. Basic Radio Channel Utilization					
Radio Type/Cache	Channel	Function	Frequency/Tone	Assignment	Remarks
King NIFC	Tac. 7	Operations		All Personnel	Training frequency
King NIFC	Disp. S	Dispatch		Command	Monitor for incidents within EFR.
King NIFC					
King NIFC					
King NIFC					
King NIFC					
King NIFC					
King NIFC					
5. Prepared by (Communications Unit)					

ICS 205 NFES 1330

FIGURE 6.3(J)

be called by the incident commander as they are needed. These individuals should be ready to act when called on. Staging areas need to be established not only for the personnel involved in the training exercise, but also for other nonfire agencies involved with the burn. The staging area should have a one-way road pattern to alleviate traffic congestion and allow for free movement of equipment and personnel. An additional review area for the media and spectators needs to be established. A live fire burn is a great publicity event, but you don't want media representatives wandering into areas where they may get injured. (Media involvement in live fire burns is discussed later in this chapter.) The instructor can count on spectators showing up on the scene of the live burn just as they do at the scene of a real incident. A training exercise like this is a perfect opportunity for the public education staff to educate people about fire and the actions of the fire department. Additionally, this person can discuss why this type of training is important.

The operations area is for those individuals who are part of the current drill. These individuals should be in full protective gear and be ready to enter the scene as directed.

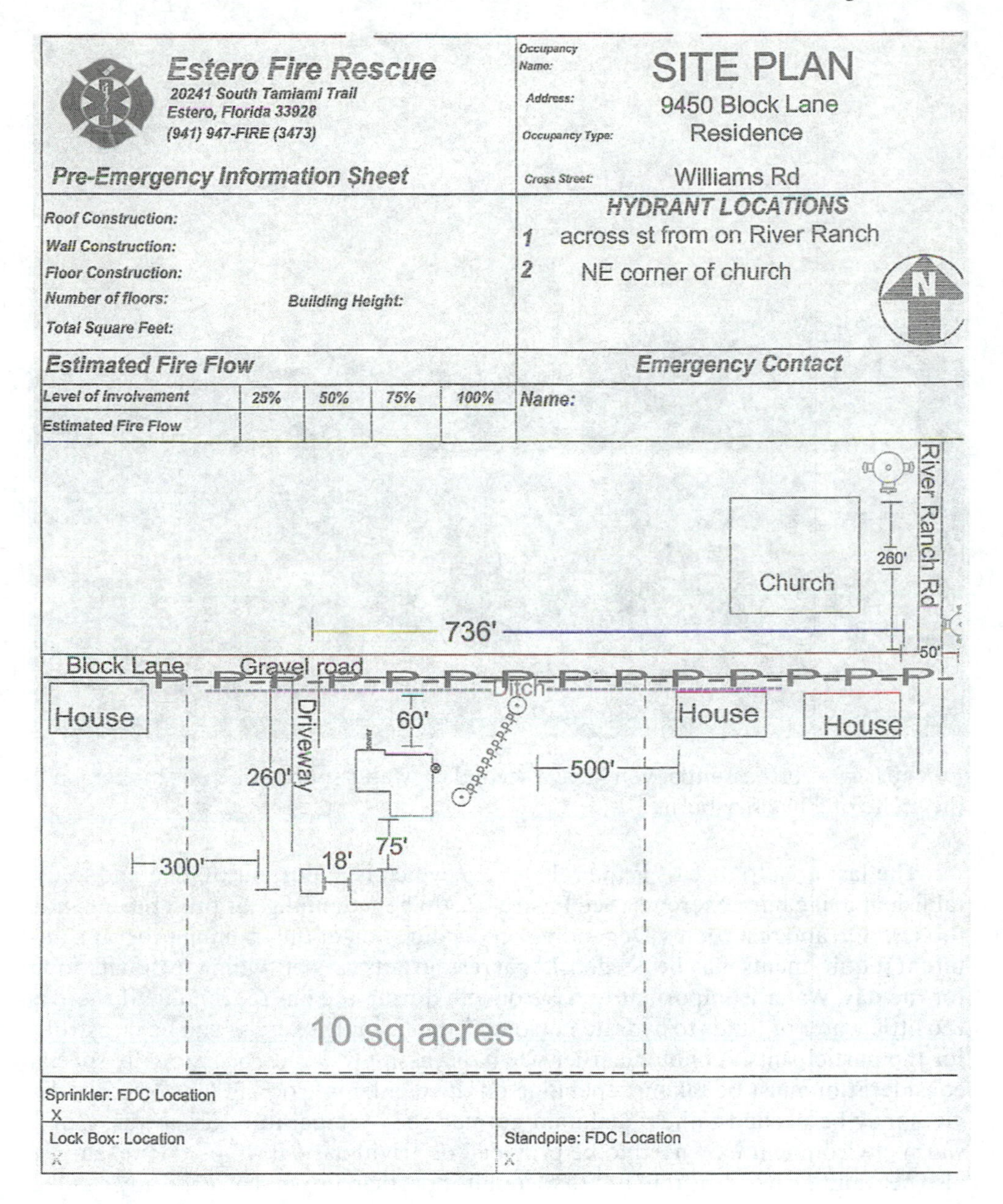

Estero Fire Rescue
20241 South Tamiami Trail
Estero, Florida 33928
(941) 947-FIRE (3473)

Pre-Emergency Information Sheet

Occupancy Name: SITE PLAN
Address: 9450 Block Lane
Occupancy Type: Residence
Cross Street: Williams Rd

HYDRANT LOCATIONS
1 across st from on River Ranch
2 NE corner of church

Roof Construction:
Wall Construction:
Floor Construction:
Number of floors: Building Height:
Total Square Feet:

Estimated Fire Flow

Level of Involvement	25%	50%	75%	100%
Estimated Fire Flow				

Emergency Contact
Name:

Sprinkler: FDC Location
X
Lock Box: Location
X
Standpipe: FDC Location
X

FIGURE 6.3(K)

The operations area should be reserved for only those who are part of the scenario. Those in the operations area would include the attack crews, Rapid Intervention Teams (RIT), instructors, safety officers, pump operators, aerial operators, and any other personnel actively involved in the exercise. The operations team should also be assigned the task of establishing the appropriate safety perimeters around the burn site.

Communications is an essential part of the training grounds, especially during a live burn. The communications center should be contacted prior to the day of the live burn to coordinate dedicated radio channel(s). (See Figure 6.5.)

FIGURE 6.4 ◆ It is essential you have a secondary water source, like this tanker, on the scene of a live fire burn.

The last area to discuss is the rehab area, which is important to plan and to establish at a live burn exercise. (See Figure 6.6.) When planning for this area, include food, water, and rest rooms. Depending on the duration of the training program, different requirements may be needed. Local restaurants may be willing to donate food for the day. Water is important to have on-site during the entire training. If there is too little water or fluids to hydrate personnel, the training exercise can be disastrous for the participants. As noted earlier when discussing weather, occasionally special consideration must be taken, depending on the elements. Cooling fans or other devices may be needed to help maintain normal body temperature on hot days, or a warm environment may need to be provided on frigid days. EMS personnel should also be standing by at the rehab area to monitor participants and provide any needed medical care.

BUILDING PREPARATION

It normally takes weeks to prepare a building for a live fire burn. Meeting the various local, state, and federal mandates plus having in place the environmental requirements needed to meet compliance does not happen overnight. The owner(s) of the building needs to be made aware of not only the requirements to burn a building, but also the amount of time it will take to prepare the structure. The time frame may not work for the owner(s), depending on the owner's schedule. In many instances these structures are being demolished for new development and the owner may need the structure demolished before a burn can be conducted.

FIGURE 6.5 ◆ Communication is an essential element of a successful live fire burn. *(Photo courtesy of James Clarke, Estero Fire Rescue)*

It takes a lot of personnel to prepare a building to burn. This is one reason that many buildings are no longer used for live fire training. As part of the preparation, the structure can be used for other training evolutions for students to learn or practice various techniques they may otherwise not have the ability to perform. It should be noted that any training conducted prior to the burn may have a detrimental effect on building preparation. Instructors need to be aware of what they can and cannot do with the building so as to not comprise the burn. In some instances the building may not be suitable for a live burn after the preburn planning inspection. In this situation the instructor may still be able to use the building for other training evolutions.

The integrity of the structure needs to be carefully evaluated before any training is conducted, and the inspection should include the services of an architect or structural engineer.

After the inspection is complete, any walls or ceilings that are found to have a highly combustible covering need to be removed. All holes in the walls or ceilings need to be patched whether they are a result of removing highly combustible materials or training evolutions. Any materials that are found to be of exceptional weight need to be removed, such as air conditioner units in the attic or on the roof of the building. Adequate ventilation openings need to be precut for each roof area. This is a great opportunity for participants to get hands-on experience in ventilation techniques.

Any utilities connected to the building must be disconnected. Locate any other live utilities on the site and clearly identify where these utilities are located on the site plan

Figure 6.6 ◆ At a live burn exercise, personnel need to go through rehab just like at a real fire incident. *(Photo courtesy of James Clarke, Estero Fire Rescue)*

and at the site. If they pose a danger to the performance, they must be shut off prior to the day of the burn if possible. Any tanks that have flammable or combustible liquids or gases, such as fuel oil or propane, need to have the appropriate precautions taken to avoid disaster. In most instances, any fuel tanks need to be removed before the day of the burn.

Any swimming pools, cesspools, or cisterns located on the site need to be filled in prior to the day of the live burn or clearly and visibly marked. (See Figure 6.7.) Satellite dishes need to be removed from the building, and if they are the large type, they should be far enough away from the structure to avoid any safety issues.

All exits need to be checked to ensure that they will open. Doors should be unlocked and operate easily. It is highly recommended that locks and latches be removed from the doors to prevent them from malfunctioning the day of the burn. Windows should be checked to make sure they open easily. In some instances it is appropriate to remove all the glass before burning the building. See Figure 6.8. If the glass is not removed, care needs to be taken for when the glass breaks during the burn. If the glass in the windows is removed, the instructor will need to board the windows for security purposes before the day of the burn, and it will be necessary on the day of the burn for

FIGURE 6.7 ◆ Hazards such as cisterns need to mark as pictured here. An orange "X" was painted on the wall closest to the hazard and the area was marked with caution tape. Personnel should also be briefed on the location of all hazards.

the windows to be closed with plywood. If the windows are boarded up, the boards need to be easily removable for safety purposes. (See Figure 6.8.)

Additionally, if there are any other building components that may affect the live burn, these need to be checked and appropriately corrected. Any fire suppression system in the building will need to be dealt with according to the needs of the burn.

When conducting the inspection, if the building has stairs, check the railings to be sure they are adequately secured. The instructor will also want to check the risers and treads to ensure adequate ingress and egress and structural integrity. If the building has a chimney, the collapse zone needs to be visibly marked so that no one enters this area. (See Figure 6.9.) This also goes for any obstacle that might fall should the building deteriorate to the point of collapse.

Any debris, either outside or inside the building, needs to be removed. If there is a porch or outside steps, the railings, treads, and decking needs to be assessed and secured. The site should also be assessed for any animals or insects that may have inhabited the area. Some animals that you should look for are snakes, rats, raccoons, alligators in some parts of the country, bats, bees or wasps, and other insects that may cause problems. In the southeast fire ants might need to be contended with when conducting outdoor training sessions.

Finally, in preparation for the live burn, the fire "sets" need to be readied. Only Class A materials can be used for live burns. No flammable or combustible liquids can be used, including igniting the combustible Class A materials. Additionally, ensure that

FIGURE 6.8 ◆ Windows may need to be removed the day of the burn. Windows should not be boarded the day of the burn and if they are the boards need to be easily removable.

no contaminated materials are in the building or on the site of the live burn. This can be accomplished through the use of the DEP or EPA checklist and approval from the state or local EPA official.

Operational Planning

The fire service instructor and the safety officer must be thoroughly informed of the building and the site prior to the live burn training. These individuals are responsible for briefing the participants before beginning any of the evolutions. All participants should take a tour through the building and around the site to be familiar with the layout, and each participant should be assigned to a company and a task to perform. The live burn training should run the same as an incident, adhering to the incident command system. All participants need to be notified of the evacuation signal for an emergency situation. The signal should be demonstrated during the plan briefing so that everyone has a clear idea of what it will sound like. (See Figure 6.10.)

Activity

You have acquired a structure for a live burn. Develop a plan for training in the building prior to conducting a live burn. Discuss any disadvantages of conducting such training in relation to conducting a live burn.

Figure 6.9 ◆ Any chimneys or antennas need to be marked. The collapse zone around the chimney should be identified with caution tape. In addition, the chimney should be marked with fluorescent paint as shown.

◆ OTHER TRAINING

As noted earlier, if NFPA 1403 requirements cannot be met to conduct a live fire burn, alternative training can be conducted using the building. For example, smoke generators can be used in the building to create a poor visibility, or blackout masks can be employed for search and rescue.

Trainer Tales

A number of years ago a bank that was going to be demolished was given to the fire agency I was working for. The building couldn't be burned since it was located in the middle of an urban environment; however, it did provide a lot of great training opportunities. In one scenario charged hose lines were placed in the building in a crisscross fashion. Firefighters were sent in one at a time with their masks blacked out. When one firefighter was partway through the building, another was sent in the structure. The firefighters were to follow the charged hose lines, identifying the male from the female coupling to determine the direction of travel to escape the building. At certain points in the drill the firefighters would encounter their hose either below or above another charged hose line. If they got off the hose line they were originally following, they would be led to a different part of the building.

In another exercise a vast amount of wire was in the building from the communication devices that had been installed in the bank. Firefighters had to follow a

FIGURE 6.10 ◆ All evolutions of the live fire burn should be planned. Planning is essential for a successful live fire burn.

charged hose line wearing blacked-out face masks—and partway through the maze, the instructors wrapped wire around the trainees' SCBA bottle. The firefighters then had to untangle themselves from the wire and continue on through the maze. Instructors and safety officers in the building were on hand at all times monitoring the progression of each firefighter.

The bank was a one-story concrete-block and stucco building. To emphasize that a structure can be turned into a variety of props, a training drill was conducted using the tele-squirts to create a standpipe system to the roof of the building. This worked well in demonstrating and practicing elevated streams, and the drill emphasized how to use a tele-squirt as a standpipe in similar scenarios. A few months later a car fire occurred on the third floor of a parking garage in the same district—and the scene went very smoothly because of the practice and drills a few weeks prior. Remember: A building can be useful for training drills even if you can't conduct a live burn.

Other training purposes for an acquired building are

- Ventilation drills
- Rescue
- Salvage and overhaul
- Aerial operations
- Forcible entry
- RIT drills
- SCBA

◆ COMMUNITY AND PARTICIPANT AWARENESS

Obviously the first step in any live fire burn is locating a structure that is slated for demolition. When the owner of the property thoroughly understands what the intention is, what is involved, and agrees to hand over the building, that clears the way for using the structure for training purposes. At this point many areas need to be considered by a variety of personnel to ensure a successful live burn at an acquired structure. Let's identify the roles of each.

Public Information Officer (PIO)

PIOs need to do what they do best: work with the instructor to contact the media as well as any others who need to be informed. The PIO should be included early on in the planning process so he/she can make the necessary preparations to notify the appropriate people. (See Figure 6.11.) The day of the burn the PIO needs to be on-site and coordinating and communicating with the media. After the burn is over, the PIO should be available to answer questions and follow up with any additional publicity about the burn. After the training, the message should convey the benefits reaped by the participants. The PIO can also be used to videotape the evolutions conducted for the post-incident analysis and future training.

Figure 6.11 ◆ The PIO should be on site in a designated area to provide information to the public when conducting a live fire burn. This is a great opportunity for media coverage. *(Photo courtesy of James Clarke, Estero Fire Rescue)*

Trainer Tales

Our agency was planning on conducting a live fire training exercise on a dead-end road. Three residents were going to be blocked in by the apparatus and hose lines, and one of the individuals expressed concern about needing to leave for a doctor's appointment. The lead instructor for the burn assured her that that wouldn't be a problem; and when the time came, provisions were made so that she could make her doctor's appointment. It's important to keep good relations with the neighbors when conducting these exercises.

Instructors

Instructors should be certified according to NFPA 1041 or requirements established by state and/or local authorities. An instructor's certification should be current and he/she should have prior experience in conducting live fire burns.

A detailed safety plan needs to be in effect that includes emergency egress, a Rapid Intervention Team (RIT), and safety officers. This is no time to take shortcuts. Everything needs to work in textbook fashion.

Safety Officers

The safety officer has the ultimate authority. One safety officer may not be enough, depending on the live burn exercise. (See Figure 6.12.) Safety officers need to be aware of all events that are occurring. Additionally, the safety officers need to make sure all participants are qualified and all safety systems are in place and followed. Safety officers entering an environment that is potentially unsafe must wear all protective gear and must not enter alone. Safety officers need to work in a team environment, the same as firefighters.

Students

A live fire burn is a great occasion to teach how fire behaves and the various conditions encountered in a real fire situation. Take advantage of this opportunity; it doesn't happen every day. As the instructor, it is your responsibility to make sure students learn correct techniques and procedures. This is an environment where you can show the students and allow them to learn the right way. Lead by example. Most of us have heard the expression "Don't do as I do, do as I say." Well, this is a great opportunity to have them do as you do—because you will do it the right way.

Other Participants

Many people envision a firefighter as someone who sits in a recliner at a fire station watching soap operas or talk shows until the alarm rings. This is an excellent opportunity to invite the community to witness firsthand what a firefighter actually does. The instructor may not want to make an open invitation to the entire community, but an invitation to the fire board or city council members is a great PR tool. The PIO or Public Education Specialist should coordinate this effort.

FIGURE 6.12 ◆ A safety officer is critical during the live burn to watch for any unsafe conditions.

Trainer Tales

When the chairperson for our fire board attended our last live fire burn at an acquired structure, he was very impressed by the actions of all the firefighters. Seeing the firefighters in action also reassured him that the millions of dollars he had just approved for new stations, new fire apparatus, and additional firefighters was not money wasted but a necessity.

◆ POST-INCIDENT ANALYSIS (PIA)

Post-incident analysis is a tool used at the conclusion of a live fire training exercise, similar to after an incident. The purpose of the PIA is to appreciate valuable lessons that were learned as a result of the observations of others during the event.

When beginning a PIA, ground rules need to be established. Participants and instructors alike need to be reminded that the PIA is for learning, not for ridicule or blame. The facilitator of the PIA needs to maintain a positive environment and keep the discussion on target for the mission that is to be accomplished. The objectives of the exercise need to be reaffirmed so that everyone remembers why they are here. Using an easel with a large pad is an excellent device to use during these analyses. The pages can be torn off and posted around the room to remind participants of items covered, such as ground rules and objectives. You can use the term "Monday-morning quarterback" in these settings—but keep in mind that such second guessing needs to be done in a positive environment. The evolution is over, and now it's time to learn what went well and what needs to be improved.

The feedback provided during the PIA needs to be an honest assessment of the operation. This is the most essential element of the analysis. Discussions on performance issues such as effective communications, tactics, and safety are just some of the concerns that should be discussed. A written report documenting the PIA needs to be completed. Aside from providing the process with a detailed review and the ability to improve the operations of the participants, this report is also a requirement in NFPA 1403.

A good format for a PIA includes the following topics:

- Scenario development
- Site planning
- Regulation compliance
- Communications
- Operations
- Staging
- Support functions
- Safety group
- Accountability
- Lessons learned
- Overall analysis

Let's examine each component in turn.

Scenario Development

When looking at this component, determine if the scenario was designed to meet the agencies' needs. Did the scenario take into account practice for multistory incidents, search and rescue, fire attack, water supply, and ventilation?

Site Planning

The following questions need to be discussed: Was the site safe? Was there an adequate prebriefing of the site for the participants? Was the construction blueprint accurate? Were there any problems with entry? What are the suggested site planning improvements?

Regulation Compliance

Ask the following question: Did the burn conform to the NFPA 1403 standard? A discussion can encompass the safety officers, instructors, rehab, and EMS as part of the dialogue. The department's Standard Operating Guidelines/Standard Operating Procedures (SOG/SOP) should also be discussed. Did any parts of the SOG/SOP not work? SOGs or SOPs are not carved in stone. The instructor may find that something within the SOG/SOP just doesn't work, and the policy needs to be changed. In the same regard, was a part of the SOG/SOP violated and does it need to be corrected in the future when conducting a live burn? Figure 6.13 shows a sample of a Live Fire Burn SOG.

Communications

This is always a biggie. Discussion should take place regarding the accuracy of the information given to the participants as to how effective it was in accomplishing the training mission. Was there any confusion of what was expected or what to do? How effective was each component of the command structure? Remember: Communication can always be improved.

Sample Standard Operating Guidelines for Live Fire Burns

Purpose: To provide procedures and guidelines for Estero Fire Rescue personnel that will be working in and around live fire training.

Scope: All Estero Fire Rescue personnel who work in and around live fire training.

Responsibility: All EFR personnel shall have a working knowledge of this procedure. All procedures shall follow NFPA 1403–14, Live Fire Training Evolutions in Structures.

Permits, Documents, Notifications, Insurance

A. Written documentation received from owner:
   1. Permission to burn structure
   2. Proof of title
   3. Certificate of insurance cancellation
   4. Water, electric, cable, gas, and all other utilities are removed
   5. Proper removal or securing of septic tank
   6. All other hazards are removed per NFPA 1403
   7. Building demolition permit from Estero Fire Rescue

B. Estero Fire Rescue burn permit received
   1. Notify the Division of Forestry (DOF)

C. Permission of owners and users of adjacent property

D. Checklist completed and notarized by homeowner

E. Letter to Department of Environmental Protection (DEP)

F. Agreement to remove debris in a timely manner after burn

Preburn Planning

A. Written preburn plan and site and building plans showing the following:
   1. Site plan drawing including all exposures
   2. Building plan including overall dimensions
   3. Floor plan detailing all rooms, hallways, and exterior openings
   4. Position of all hose lines, including backup lines
   5. Location of emergency escape routes

B. Available water supply determined

C. Backup water supply

D. Required fire flow determined for fire building and exposures

E. Required reserve flow determined (50 percent of fire flow)

F. Operations area established and perimeter marked

G. Early notification of law enforcement for traffic control

H. Notify utility company of additional water demand

(*Continued*)

FIGURE 6.13

Building Preparation

A. Building inspected to determine structural integrity
  1. Call for inspection from Estero Fire Rescue
B. All utilities disconnected
C. Materials of exceptional weight removed from above training area
D. Ventilation opening precut
E. Chimney checked for safety
F. Cisterns, wells, cesspools, and other ground openings fenced or filled
G. All extraordinary exterior and interior hazards removed
H. Class A materials only
I. No flammable or combustible liquids used in live fire training evolutions
J. Stairwells and porches secured/checked for safety

Preburn Procedures

A. All participants briefed regarding:
  1. Building layout
  2. Crew and instructor assignments
  3. Safety rules
  4. Building construction
  5. Objectives of drill
B. All equipment in proper working order
C. Written objective and methodology of burn
D. Possible structure cuts to aid in controlling demolition
E. ICS/IMS shall be utilized
F. All involved personnel shall utilize full protective clothing and equipment
G. Rehab established
H. Ignition officer

Postburn Procedures

A. All personnel accounted for
B. Remaining fires overhauled
C. Training critique conducted
D. Training records completed
E. Property turned back over to owner
F. Call for inspection from Estero Fire Rescue

Instructor

A. Plan and coordinate all training activities
B. Monitor activities to ensure safety
C. Inspect building integrity prior to each burn
D. Assign safety and instructors as needed

(*Continued*)

FIGURE 6.13

Safety Officer

A. Prevent unsafe acts
B. Eliminate unsafe conditions
C. Coordinate lighting of fires with instructor
D. Ensure compliance of participants personnel equipment to applicable standards
E. Ensure that all participants are accounted for both before and after each evolution

FIGURE 6.13 ◆ *(Concluded)*

Operations

One aspect to discuss regarding operations is determining if all components of the Incident Management System were utilized. This would include if division/groups were established. Was an action plan developed? Were proper tactics and strategies implemented? This would include such topics as: search and rescue, fire suppression, and ventilation.

Staging

Questions to be asked include: Was the staging location readily identified? Was there ease of accessibility for all responding apparatus? Was there proper traffic control? (See Figure 6.14.)

FIGURE 6.14 ◆ A designated staging area should be identified. This is also a great area to brief all the participants about the drill.

Support Functions

Discuss whether support functions were properly identified and implemented. This would include: rehabilitation group, proper hydration, medical monitoring, air supply, and coordination with outside agencies.

Safety Group

Ask the following: Were there an adequate number of safety officers? Were safety officers properly identified? Were Rapid Intervention Teams established? Were all personnel on the fire ground wearing proper PPE? Did all companies operate in a safe manner?

Accountability

Were all company Personnel Accountability Recorder (PAR) tags given to the accountability officer? Were PAR reports conducted in accordance with SOPs? (See Figure 6.15.)

Lessons Learned

During this aspect of the review discuss the SOPs that are related to fire ground tactics. This is also a good time to discuss future training required for specific topics. Discussion on improving fire ground communications is always an item of interest during lessons learned. This would also be an excellent time to review multicompany operations.

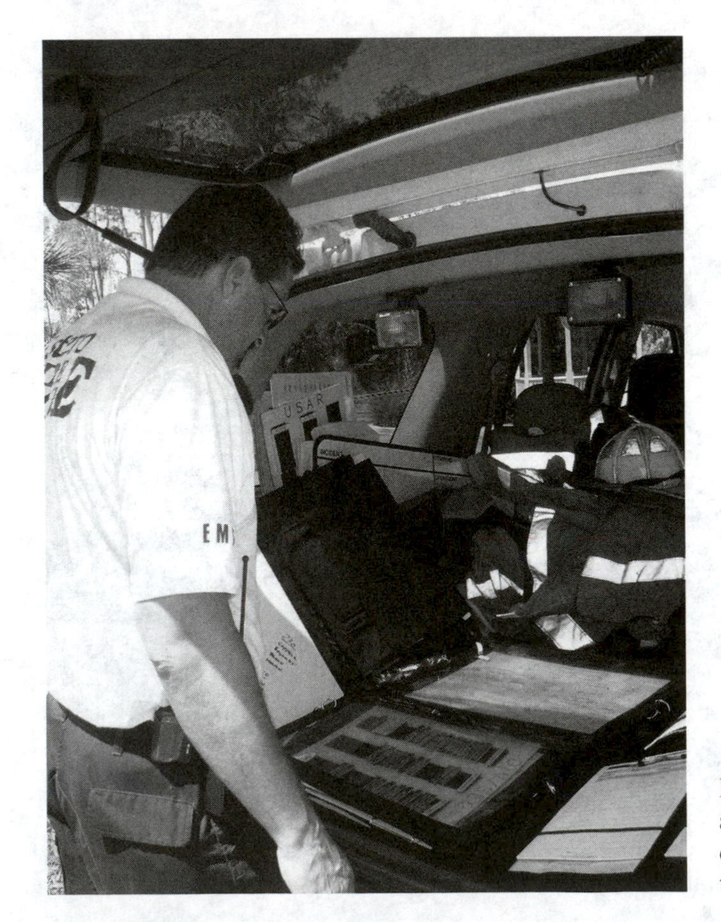

FIGURE 6.15 ◆ Accountability of personnel is essential to provide for the safety of personnel.

Overall Analysis

Here are some questions that can be discussed dealing with the overall analysis of the training: Was this exercise beneficial? What improvements can be incorporated? What are your future expectations for live burn training?

◆ TYPES OF BURN BUILDINGS AND PROPS

There are different types of burn buildings. (See Figure 6.16.) This chapter has mostly discussed acquired structures; however, buildings are constructed for repeat burnings. These buildings are typically fueled by Class A materials or some type of gas. Fuel-simulated fires are becoming more popular for training using live fire. These simulators have advantages and disadvantages.

The primary advantage of a gas-fueled simulator is that they are considerably safer than using an acquired structure or even Class A burn materials in a burn building. (See Figure 6.17.) Fuel-supplied fire simulators have a number of built-in safety features. Automatic shut-off valves control the heat and smoke, and the rooms can be monitored. Remote video cameras can be placed in the burn rooms to capture video for training purposes, while at the same time personnel can monitor what is happening in the room. (See Figure 6.18.)

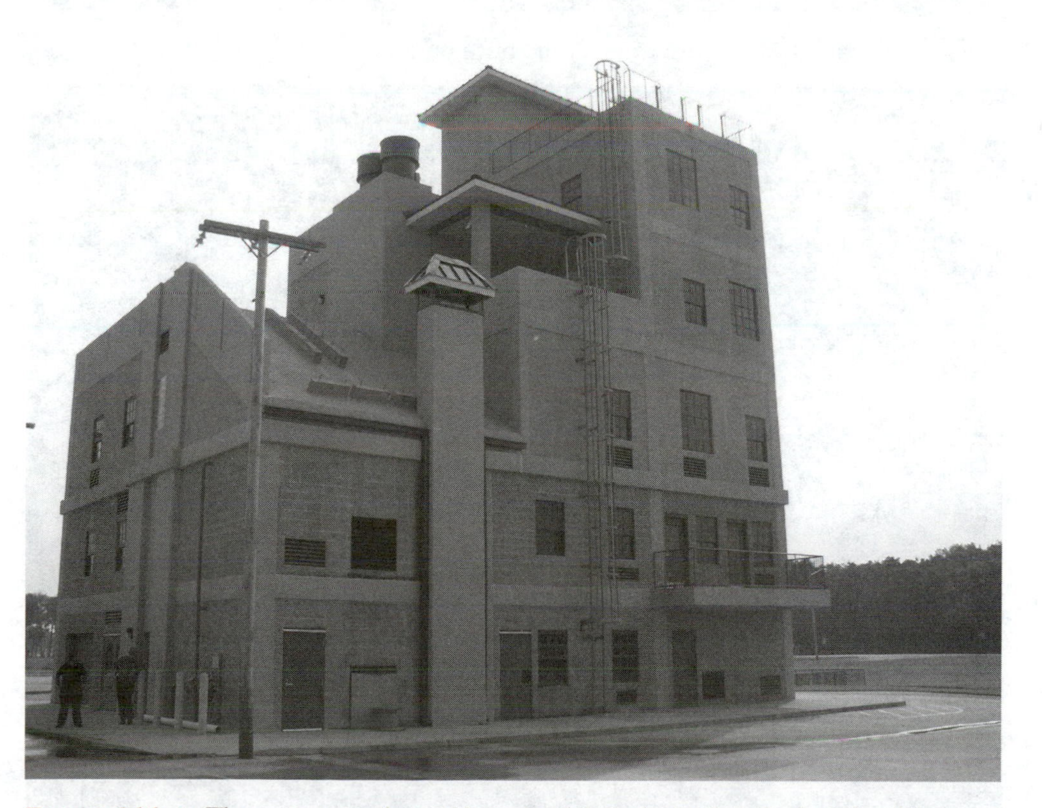

FIGURE 6.16 ◆ There are a variety of burn buildings, including the burn building and tower at the Tennessee Fire Training Center.

FIGURE 6.17 ◆ A Class A fire room at the burn building

FIGURE 6.18 ◆ A gas-powered stove fire simulator

FIGURE 6.19 ◆ Some burn buildings have cameras placed in the rooms to monitor the personnel and capture video for future training during the PIA.

One disadvantage of a simulated fire fueled by propane is it will not give you the same effect as a real fire scenario; however, such fires can be used to teach technique and attack. Another disadvantage of simulator training using fuel-generated fire is that typically the props can't be moved. (Some facilities have taken this into consideration and designed their props to be mobile.) In instances where the props are stationary, each time the person enters the room, they will find the prop or the fire in the same location. A good instructor will be creative and overcome this obstacle by designing different evolutions for the participants. They may enter the room from one direction the first time, but the next time they may have to approach from an entirely different position.

Figures 6.19 through 6.21 illustrate some of the other burn buildings available to conduct live fire training.

Exterior Props

Exterior props include large- and small-scale fires that can be put out with a fire extinguisher or can be used for team hose handling and hazardous-materials response training. There are props designed to simulate a fire under, around, and on top of a rail car, tank truck, or stationary facility. Burn pans can be used to demonstrate and practice using fire extinguishers. Figures 6.22 through 6.25 shows simulated situations.

FIGURE 6.20 ◆ A gas-powered storage fire simulation

FIGURE 6.21 ◆ A flashover gas-powered fire simulator

FIGURE 6.22 ◆ A simulated swift water scenario from the Tarrant County Training Center

FIGURE 6.23 ◆ A simulated rail car derailment with a fuel spill. The fire is from a fuel source and can be controlled by the instructor.

FIGURE 6.24 ◆ A flange gas-powered simulator

FIGURE 6.25 ◆ Two firefighters practice extinguishing an engine compartment fire on a simulated gas-powered simulator of a vehicle.

◆ SUMMARY

It is becoming more and more difficult to conduct live burns. A number of issues may preclude or deter the use of structures to conduct live fire training, the issues being the number of injuries and fatalities as a result of live burns and environmental factors. Hence, using fire simulators is becoming more prevalent in fire training. This chapter has looked at the legal issues, discussed how to design and plan a live burn, and described a number of simulation props available for the fire service to conduct live fire training. It is essential that the fire service instructor provides a safe training environment when conducting live burns.

Review Questions

1. Which NFPA standard deals with live burns?
2. Identify the agencies you would need to deal with if conducting a live burn in your jurisdiction.
3. Describe the process of conducting a live fire burn on an acquired structure.
4. Describe the steps to establish a live fire burn.
5. Describe the safety procedures you need to take at a live burn.
6. Define and describe the key elements of a PIA.
7. Describe the documentation you need to conduct a live fire burn at an acquired structure.
8. List at least four interior simulator props.
9. List at least four exterior props.

References

National Fire Protection Agency. 2002. *NFPA 1403, standard on live fire training evolutions, 2002 ed*. Batterymarch Park, MA.

Bibliography

Montgomery County Fire Rescue, Maryland. 2002. *Post incident analysis*. Retrieved on February 10, 2004, from *http://www.co.mo.md.us/services/dfrs/frc/frcpolicy/frcp2002attachment1.PDF*.

Cannell, J., Reall, J., and Bernzweig, D. *Hi-rise fire training: Operation "H.O.T.T." Fire Engineering.com*. Retrieved on January 14, 2003, from *http://fe.pennnet.com/Articles/Article_Display.cfm?Section=Archives&Subsection=Display&ARTICLE_ID=118240&KEYWORD=cannell.*

Thiel, A., Stern, J., Kimball, J., and Hankin, N. 2003. *Trends and hazards in firefighter training*. (No. USFA-TR-100.) Emmitsburg, MD: Federal Emergency Management Agency.

CHAPTER 7

Testing and Evaluation Techniques

Terminal Objective

The participant will be able to construct, administer, and evaluate an assessment and testing instrument.

Enabling Objectives

- Define the four levels of evaluation according to Kirkpatrick.
- Differentiate between formative and summative evaluation.
- Differentiate between informal and formal evaluation.
- Define a norm-referenced test as compared to a criterion-reference test.
- Define the different type of tests.
- Construct a test using a test blueprint.
- Discuss the difference of the various types of tests.
- Conduct an analysis of test questions.
- Determine how to conduct reliability and validity of test scores.
- List various sources for tests.

JPR NFPA 1041—Instructor I

4-5.2 Administer oral, written, and performance tests, given the lesson plan, evaluation instruments, and the evaluation procedures of the agency, so that the testing is conducted according to procedures and the security of the materials is maintained.

4-5.3 Grade student oral, written, or performance tests, given class answer sheets or skills checklists and appropriate answer keys, so that the examinations are accurately graded and properly secured.

4-5.4 Report test results, given a set of test answer sheets or skills checklists, a report form and policies and procedures for reporting, so that the results are accurately recorded, the forms are forwarded according to procedure, and unusual circumstances are reported.

JPR NFPA 1041—Instructor II

5-2.6 Evaluate instructors, given an evaluation form, department policy, and job performance requirements, so that the evaluation identifies areas of strengths and weaknesses, recommends changes in instructional style and communication methods, and provides opportunity for instructor feedback to the evaluator.

5-5.2 Develop student evaluation instruments, given learning objectives, audience characteristics, and training goals, so that the evaluation instrument determines if the student has achieved the learning objectives, the instrument evaluates performance in an objective, reliable, and verifiable manner, and the evaluation instrument is bias-free to any audience or group.

5-5.3 Develop a class-evaluation instrument, given agency policy and evaluation goals, so that students have the ability to provide feedback to the instructor on instructional methods, communication techniques, learning environment, course content, and student materials.

5-5.4 Analyze student evaluation instruments, given test data, objectives and agency policies, so that validity is determined and necessary changes are accomplished.

JPR NFPA 1041—Instructor III

6-2.5 Construct a performance-based instructor evaluation plan, given agency policies and procedures and job requirements, so that instructors are evaluated at regular intervals, following agency policies.

6-2.7 Present evaluation findings, conclusions, and recommendations to agency administrator, given data summaries and target audience, so that recommendations are unbiased, supported, and reflect agency goals, policies, and procedures.

6-5.2 Develop a system for the acquisition, storage, and dissemination of evaluation results, given agency goals and policies, so that the goals are supported and those impacted by the information receive feedback consistent with agency policies, federal, state, and local laws.

6-5.3 Develop a course evaluation plan, given course objectives and agency policies, so that objectives are measured and agency policies are followed.

6-5.4 Create a program evaluation plan, given agency policies and procedures, so that instructors, course components, and facilities are evaluated and student input is obtained for course improvement.

◆ INTRODUCTION

As part of the learning process, the instructor needs a mechanism to evaluate the student and identify whether the student is achieving the objectives and goals of instruction. (See Figure 7.1.) It is not only a measurement of the student's progress but an indicator to make sure the instructor is covering the material effectively. The evaluation

FIGURE 7.1 ◆ Students can be tested and evaluated using many different techniques.

and testing process helps determine strengths and weaknesses of the students as well as the program.

Program evaluations help the instructor improve the quality of instruction. New instructors may not design and develop test questions, but they should at least have familiarity with the concepts of evaluation and how to construct solid tests. This is the case even if using a prepared test bank. The instructor should understand how to determine if these questions are well written and match the objectives to the lesson plan and standard.

◆ KIRKPATRICK'S FOUR LEVELS OF EVALUATION

According to educator Donald Kirkpatrick, there are four levels of evaluation:

Level I—Reaction
Level II—Learning
Level III—Transfer
Level IV—Business Results

Let's examine these in turn.

LEVEL I: REACTION

Level I assesses the students' initial reactions to a course. This, in turn, offers insights into their satisfaction with a course, a perception of the value of the class. Instructors usually assess this aspect through an evaluation form. Occasionally instructors use focus groups and similar methods to receive more specific comments—called "qualitative feedback"—on the courses, although this is not typically the norm. Reaction assessment could be as simple as asking the students at the end of the class to discuss what they thought of the class and how they would make the course better. It is probably safe to say that almost 100 percent of instructors conduct a "Level I" evaluation at the end of class as required by most academic institutions. (See Figure 7.2.)

Program Evaluation Form

Program Title: ______________________ Training Location: ______________________

Participant Name (optional): ____________________________ Date: ________________

Instructor's Name: ______________________

Course Content (Circle your response to each item.)

1 = Strongly disagree 2 = Disagree 3 = Neither agree/nor disagree 4 = Agree
5 = Strongly agree

1. I was aware of the prerequisites for this course.	1	2	3	4	5
2. I had the prerequisite knowledge and skills for this course.	1	2	3	4	5
3. I was well informed about the objectives of this course.	1	2	3	4	5
4. This course met my expectations.	1	2	3	4	5
5. The content is relevant to my job.	1	2	3	4	5

Course Design (Circle your response to each item.)

6. The course objectives are clear to me.	1	2	3	4	5
7. The course activities stimulated my learning.	1	2	3	4	5
8. Interactive multimedia was essential in the course.	1	2	3	4	5
9. The activities in this course gave me sufficient practice and feedback.	1	2	3	4	5
10. The test(s) in this course were accurate and fair.	1	2	3	4	5
11. The difficulty level of this course is appropriate.	1	2	3	4	5
12. The pace of this course is appropriate.	1	2	3	4	5

Course Instructor (Facilitator) (Circle your response to each item.)

13. The instructor was well prepared.	1	2	3	4	5
14. The instructor was helpful.	1	2	3	4	5

Course Environment (Circle your response to each item.)

15. The training facility at this site was comfortable.	1	2	3	4	5
16. The training facility at this site provided everything I needed to learn.	1	2	3	4	5

Course Results (Circle your response to each item.)

17. I accomplished the objectives of this course.	1	2	3	4	5
18. I will be able to use what I learned in this course.	1	2	3	4	5
19. Video is an important aspect of the course.	1	2	3	4	5

(Continued)

FIGURE 7.2 ◆ A sample course evaluation

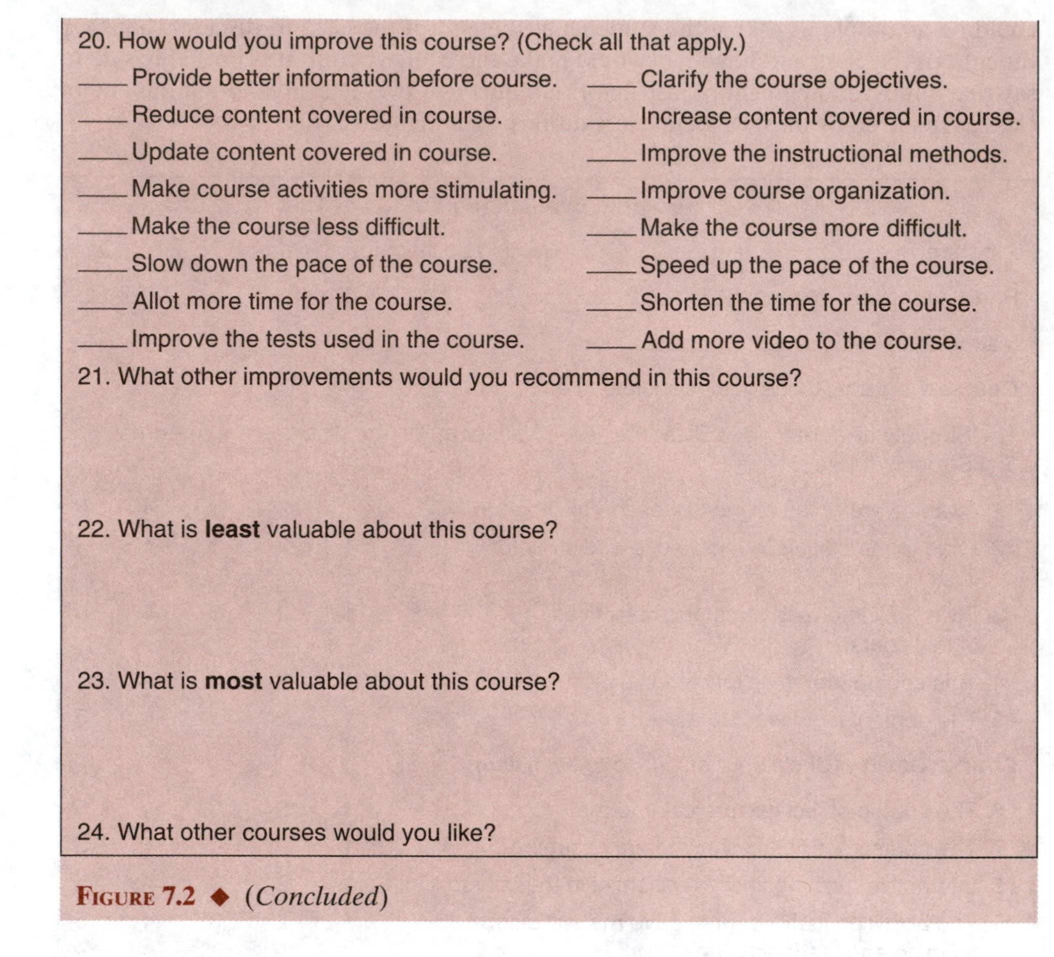

20. How would you improve this course? (Check all that apply.)

____ Provide better information before course.	____ Clarify the course objectives.
____ Reduce content covered in course.	____ Increase content covered in course.
____ Update content covered in course.	____ Improve the instructional methods.
____ Make course activities more stimulating.	____ Improve course organization.
____ Make the course less difficult.	____ Make the course more difficult.
____ Slow down the pace of the course.	____ Speed up the pace of the course.
____ Allot more time for the course.	____ Shorten the time for the course.
____ Improve the tests used in the course.	____ Add more video to the course.

21. What other improvements would you recommend in this course?

22. What is **least** valuable about this course?

23. What is **most** valuable about this course?

24. What other courses would you like?

Figure 7.2 ◆ *(Concluded)*

LEVEL II: LEARNING

At this level the instructor assesses the amount of information students have learned. Kirkpatrick defines this level as determining if the objectives have been mastered via some demonstrable means such as an oral, written, or skills test. The criteria used for the testing should be the objectives established for the course that explicitly state the skills that participants should be able to perform after taking a course. Because the objectives are the course requirements, a Level II evaluation assesses conformance to requirements or quality. In many instances this test is performed in the class.

LEVEL III: TRANSFER

Level III assesses how much of the material the student retains from the classroom and actually uses in the work environment. This assessment typically occurs anywhere from six weeks to six months (perhaps longer) after taking the course and is based on tests, observations, surveys, and interviews with coworkers and supervisors. Like the Level II evaluation, Level III assesses the requirements of the course and can be

FIGURE 7.3 ◆ Evaluation should not stop after the student leaves the classroom. Retention of the information and return on investment need to be evaluated throughout time after the course.

viewed as a follow-up evaluation of quality. (See Figure 7.3.) This is not done on a regular basis in the fire service, which has not been active in determining if what is learned in the training environment is retained and used in the work environment over time.

LEVEL IV: BUSINESS RESULTS

At this level the instructor assesses the financial impact of the training course on the bottom line or the return on investment to the organization six months to two years after the course (the actual time depends on the context of the course). The fire service has been very delinquent in determining these results. The instructor needs to ask the question: Is what we are doing making sense? Is it cost-effective to continue to train and do the things we do? It's easy to answer yes; but unless instructors actually begin scientifically evaluating this information, there is nothing on which to base the findings.

Level IV is the most difficult to measure. First, most training courses don't have explicitly written business objectives, such as "This course should reduce support expenses by 20 percent." Second, the methodology for assessing business impact is not yet refined—some assess this by tracking business measurements, others by observations, some by surveys, and still others assess by qualitative measures. Last, after six months or more, evaluators have difficulty attributing altered business results solely to training when changes in personnel, systems, and other factors might also have contributed to performance.

INSTRUCTORS' USE OF THE KIRKPATRICK MODEL

Instructors follow this model at different levels. As noted previously, Levels I and II are probably conducted more frequently than Levels III and IV in the fire service. It's time instructors conduct more Level III and IV evaluations. During the budget process the instructor typically asks for additional money for the training program. Using Levels III and IV to defend the reasoning can help produce the needed documentation and proof that the training works and potentially get the funding to continue and/or improve the training program. The following information covers the various levels of assessment.

EVALUATION

An evaluation is the process of making a value judgment based on information from one or more sources. In classroom evaluation, it is a mechanism of determining student progress toward, or the attainment of, stated cognitive, psychomotor, and affective objectives. The evaluation process should look at two components:

1. Instruction as provided by the teacher
2. The performance of the student on course and lesson objectives

PURPOSE OF EVALUATION

Evaluations in the classroom are important for a variety of reasons. First, evaluations provide feedback to students on their progress or performance. Second, evaluations provide some students with gratification and motivation to succeed. Third, evaluations measure the effectiveness of the instructor's teaching style and the lesson content. Finally, evaluations measure the effectiveness of the educational program in meeting the written goals and objectives established for the program. The two types of evaluation are formative and summative.

Formative Evaluation

Formative evaluation is the ongoing evaluation of the students and instruction, conducted throughout the course, to change or adapt the program accordingly. The formative evaluation process compares the overall goal of instruction, lesson objectives, and the content to the students' performance. Formative evaluation also compares the objectives of the course to the testing strategy of the course. Formative evaluation typically occurs during the development of the course or test. Pilot testing is considered formative evaluation.

The instructor can gain valuable early insight into the program by using formative evaluation. In many instances instructors teaching over multiple weeks will give a feedback evaluation halfway through the program instead of getting feedback only at the end of the class. This allows instructors to take the information they get from the feedback and alter the teaching style, provide remediation, or change whatever needs to be changed to accommodate the needs of the students.

There are a variety of methods of performing formative evaluation. The module or section testing conducted during the course is formative evaluation. Indeed, anything that helps to gauge the progress of the students' learning and lets the instructor

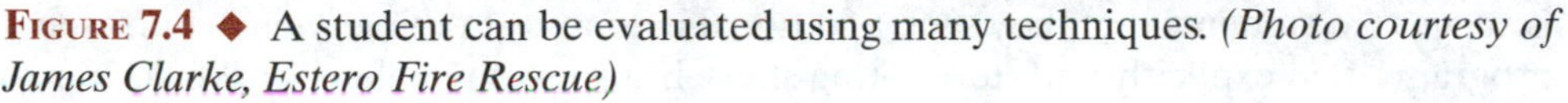

Figure 7.4 ◆ A student can be evaluated using many techniques. *(Photo courtesy of James Clarke, Estero Fire Rescue)*

use the information to alter the course delivery is a process of formative evaluation. (See Figure 7.4.)

Summative Evaluation

Summative evaluation is typically performed at the end of a course or program, module, or series of modules. This type of evaluation provides feedback to the students on their mastery of the content of the course material. It also gives instructors feedback on the effectiveness of their teaching strategy and the ability to improve their teaching performance for the next course.

There are various methods of performing summative evaluation during a course or class. The first—and most typically used—is an evaluation form given to each student when the course is complete to rate the course, the instructor, and the room.

Final exams are another form of summative evaluation. These written or practical tests aid in determining if the goals and course outcomes were met. It is important to make sure each test item is validated. (Test validation will be discussed later in this chapter.)

A summative evaluation can also be formative, depending on the context in which it's used. For example, a multiple choice final exam given at the end of a topic will be both formative and summative. It is summative because it represents the end of that topic area; it is formative because it can be used to change future courses that an instructor teaches.

FORMAL AND INFORMAL EVALUATION

Both formal and informal strategies are critical to the success of courses and programs. Some of the evaluation strategies listed can be conducted formally or informally. Let's take a look at both types of evaluations.

FORMAL EVALUATION

Formal evaluation—typically referred to as a test or an assessment—is used to assess the student's attainment of interim and/or final course objectives. A formal written examination can determine a grade for a course or serve as a means for the student to continue in the course or program. By reviewing the test and allowing students to challenge questions, it can serve as a powerful learning tool. However, for students to successfully challenge a question, they will need to prove that their answers are correct by using textbooks, class notes, and so on to prove why the answer is correct.

There are problems associated with formal evaluation. First and foremost, it places stress on the student—especially an ill-prepared student. Test-taking anxiety is very common, and there is someone in just about every class who suffers from test anxiety. The instructor will also find students who have disabilities that may interfere with their test-taking ability. Although the instructor can't diagnose students with a learning disability, if a student tells the instructor that they have such a disability, the instructor needs to provide reasonable accommodations. One thing an instructor can do is suggest to students that they seek additional help in test taking. Many courses and programs deal explicitly with test-taking skills. In fact, many colleges have such programs on campus for students at no additional cost, and, of course, a number of similar programs are available in the private sector.

Formal testing is only useful *during* the course or program. When it comes down to the end of the course, and the final exam is administered, it is essentially too late to offer any remediation to a student who has not kept up with or retained the work and that student may fail.

Needless to say, if the students' successful completion of a course is based on passing a formal written test at the end of the course, it is not fair to them if the final exam is the first time they have been tested on the material. The best way for students to be prepared to take a comprehensive test is to be tested in smaller chunks throughout the course.

Trainer's Tip

I like to give a quiz on each chapter, a mastery test on multiple chapters, and then a final comprehensive test at the end of the course. By the time the student reaches the point of taking the certifying exam, he/she has seen thousands of test questions and can read and answer a question successfully.

INFORMAL EVALUATION

An informal evaluation—which doesn't necessarily result in a grade or the grade may not be recorded—is a less structured method of assessing student achievement. This

Figure 7.5 ◆ As an instructor you need to give timely feedback to the student. *(Photo courtesy of James Clarke, Estero Fire Rescue)*

type of evaluation is primarily used to provide corrective feedback to both student and instructor. There are many benefits to the student using this type of evaluation. One is that the student will identify his/her weakness along with strengths on a particular topic. This allows the instructor to give suggestions for improvement. It may also serve as a wake-up call if the student realizes how much more study or work is needed on a certain subject. The instructor also benefits by comparing results from the class to identify trends and problems. This will help in developing corrective instruction or remediation.

Instructors need to be cautious in how they present an informal evaluation. It may cause conflict when students have the expectation of a formal evaluation or test. Students may not perceive the value in informal evaluation techniques because grades are not recorded, so the instructor may not get an adequate indicator of how he/she is actually doing. In fact, instead of spending class time doing informal evaluations, they may work better as take-home assignments. If the instructor doesn't grade or provide feedback on informal evaluations, these assignments may be diminished in importance in the students' eyes. (See Figure 7.5.)

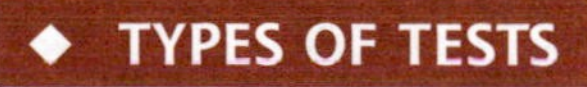

◆ TYPES OF TESTS

Tests may be written, oral, or practical in form. The next section is devoted to looking at the various testing modes.

Criterion- versus Norm-Referenced Testing

The first issue is to differentiate between a criterion-referenced and a norm-referenced test. The following chart illustrates the differences between these two types of tests.

	Criterion-Referenced Tests	*Norm-Referenced Tests*
Purpose	This type of test determines whether each student has achieved specific skills or concepts. It illustrates how much students know before instruction begins and after it has finished.	This type of test ranks each student with respect to the achievement of others in broad areas of knowledge. It discriminates between high and low achievers.
Content	Measures specific skills that make up a designated curriculum. These skills are identified by teachers and curriculum experts. Each skill is expressed as an instructional objective.	It measures broad skill areas sampled from a variety of textbooks, syllabi, and the judgments of curriculum experts.
Item Characteristics	Each skill is tested by at least four items to obtain an adequate sample of student performance and minimize the effect of guessing. The items that test any given skill are parallel in difficulty.	Each skill is usually tested by less than four items. Items vary in difficulty. Items are selected that discriminate between high and low achievers.
Score Interpretation	Each individual is compared with a preset standard for acceptable achievement. The performance of other examinees is irrelevant. A student's score is usually expressed as a percentage. Student achievement is reported for individual skills.	Each individual is compared with other examinees and assigned a score—usually expressed as a percentile, a grade equivalent score, or a stanine. Student achievement is reported for broad skill areas, although some norm-referenced tests do report student achievement for individual skills.

Source: Huitt, W. (1996). Measurement and evaluation: Criterion- versus norm-referenced testing. *Educational Psychology Interactive.* Valdosta, GA: Valdosta State University. Retrieved September 2004, from *http://chiron.valdosta.edu/whuitt/col/measeval/measeval.html.*

WRITTEN EXAMINATIONS

There are five types of written exams:

1. Multiple choice
2. True/false
3. Matching
4. Completion or fill in the blank
5. Essay

Before the types of tests are examined, the source of test items needs to be examined.

SOURCE OF TEST ITEMS

The two main sources for test items are course content and lesson objectives. Whether instructors construct the test or the test is supplied for the course, they need to make sure the materials are covered. Students will try to get the instructor to throw questions out because the question was never covered in class. The materials on the test don't all have to have been covered in class, but the information should come from reading assignments, projects, or some other form of independent study. Also the instructor needs to inform students what they will be responsible for in addition to what is covered in the classroom. If the instructor tells students that they are responsible for any materials handed out in class, covered in class, and covered in the text, it all becomes fair game for the test. If instructors tell students that only the material covered in class will be tested, they should not include items from the text because this will harm their integrity.

The advantages of written examinations are several. They can be used with a large number of students, and it is easy to measure cognitive objectives. (See Figure 7.6.) They typically provide for consistent scoring, and it's easier and quicker to grade and compile the results than for other types of examinations. The exception to this is the essay and the short answer exam that take much longer to prepare and are more difficult to grade.

There are also disadvantages to written exams. They are time consuming to develop, and as you climb to the higher levels of learning, it becomes more difficult to develop adequate measurements. They are also complex to validate—and the instructor should not develop a test and then administer it without formative validation. Pilot testing an exam is a formative evaluation process that will ensure the validity and reliability of the exam's test scores.

Another disadvantage to written exams is that they can discriminate against students with reading difficulties. Poorly written responses may evaluate a student's reading ability more than they evaluate knowledge of the material.

Finally, fire fighting requires individuals to be proficient at particular skills that written examinations cannot measure. The instructor can ask questions about the procedure to perform, but the actual skill demonstration can't be evaluated via this type of test.

TEST CONSTRUCTION TECHNIQUES

Constructing a test is as much a process as developing a course curriculum or preparing a lesson plan to teach a course.

FIGURE 7.6 ◆ The written exam is the most common form of testing and evaluation. It can be given to a large number of students at one time and is typically the easiest test to conduct.

GENERAL GUIDELINES FOR WRITTEN TEST ITEM CONSTRUCTION

A test must be related to objectives and developed from a blueprint that's an outline for the test. The blueprint should include test items on each objective in each chapter being tested. The blueprint helps to decide the level of coverage for each item.

Figure 7.7 is an example of a completed test blueprint for a 40-question exam for this chapter, using a select number of objectives. The first step in completing a test blueprint is to list the objectives the test will measure. The next step is to determine how many questions at each level of Bloom's taxonomy you want to include. The total number of questions for each objective is listed in the far right column. Note that the higher the level of Bloom's, the harder to write a test question. Most written test questions are at the knowledge and comprehension level.

Once the number of questions is determined, there is a blueprint for how many questions to assess at each level for each objective.

When the test blueprint is complete, the next step is to determine the appropriate type of exam to construct. Considering the domain of learning and the level of the domain (high or low level or level 1, 2, or 3) will help to determine the appropriate format to use. The following is a partial list of suggestions to follow:

- *Low-level cognitive:* multiple choice, matching, true/false, simple completion (fill-in-the-blank) or short answer essay, and oral exam

Objective	Knowledge	Comprehension	Application	Analysis	Synthesis	Evaluation	Total
Define the four levels of evaluation according to Kirkpatrick.	3	3	2				8
Differentiate between formative and summative evaluation.	3	3	3				9
Differentiate between informal and formal evaluation.	1	1					2
Define a norm-referenced test as compared to a criterion-referenced test.	2	2	2				6
Define the different types of tests.	3						3
Construct a test using a test blueprint.			3		2		5
Conduct an analysis of test questions.	2	2	1	1	1	1	8

FIGURE 7.7 ◆ An example of a test blueprint.

- *High-level cognitive:* long and short answer essay, fill in the blank, some true/false and completion, oral exams, projects (case studies, for example), and observational reports
- *Low-level psychomotor:* rote skills, oral, and observational reports
- *High-level psychomotor:* situational scenarios, projects (designing scenarios, for example), and observational reports
- *Low-level affective:* oral, short answer essay, projects (opinion papers, for example), and observational reports
- *High-level affective:* oral and situational scenarios, projects (group-designed presentations, for example), and observational reports

Now the instructor needs to organize the exam in a logical manner. It's best to group like items (similar content area) together on a written exam. Professional test developers recommend that mixing questions of various topics is not the appropriate way to construct a test. The questions need to follow a linear or logical sequence as in an oral or scenario-based exam.

TEST TIMING

In instances when a test or part of a test is timed, allow an appropriate amount of time to answer questions or perform a skill. This can be determined by having six to eight individuals who are proficient at that particular skill or duty perform it and time them. By averaging out the completion times, an instructor can establish a parameter for the students' performance.

To prepare the students for a licensure or certification exam, determine the time allotted for the test and then mirror the timing strategy of the exam in the preparatory exam. Timing tests may also be appropriate if an instructor is trying to get students to think fast. Fire fighting requires individuals to think on their feet and make quick, critical decisions. Testing this ability helps students learn the pressures they'll be facing in the real world.

The following are some suggested timing strategies to incorporate in testing: For a standard multiple choice test, allow one minute per item. If the question includes a scenario, allow two to three minutes to read the scenario, then one minute for each multiple choice question. If the student has to compose an essay question, allow time to read a scenario and then formulate and write the answer. If the instructor is creating a situational skill rather than a rote one, allow more time to respond. A good rule is the higher the level of domain of the question, the more time an instructor should allow for the students' response.

ADDITIONAL ITEMS TO CONSIDER

The directions for the test need to be clear and complete. For example, can students write on the test? If they need to use a pencil, make sure they are aware of this requirement prior to test time. Also, make sure students know if there is a time limit and whether or not breaks will be allowed during the test.

It's a good idea to have another instructor review the examination for clarity and completeness. Make sure that the exam is legible and free of typographical, grammatical, spelling, and content errors. This can be done during the formative evaluation of the test. The instructor won't look good if the test is full of errors. This may also distract and confuse the test taker. The graphic design and fonts need to be consistent—use either all capitals or all lowercase letters throughout the test for the key and the distracters on the test. This is true for enumerating (a, b, c, d or A, B, C, D) as well as for the first word of the key and the distracter. (Distracters are discussed in more depth later in this chapter.)

Using consistent punctuation is another factor in writing tests. Along with this goes consistency regarding underlines, boldface, or italics to accentuate material in the test.

The instructor needs to position key and distracters appropriately. Watch for answers that build logically. If the answer choices are the letters a, b, c, and d, place them in that order. Place numbered answers in ascending or descending order.

Trainer's Tip

Good Test Question Formats

The following are examples of test question formats to be used. Lowercase letters with answers in one-column list form:

1. Which NFPA standard addresses personal protective clothing for wildland fire fighting?
   a. 1971
   b. 1973
   c. 1975
   d. 1977
2. According to NFPA 1581, how often should personal protective clothing be cleaned and dried?
   a. Once every month
   b. At least every two months
   c. At least every six months
   d. Once per year

Lowercase letters for a two-column format:

1. Which NFPA standard addresses personal protective clothing for wildland fire fighting?

a. 1971	c. 1975
b. 1973	d. 1977

2. According to NFPA 1581, how often should personal protective clothing be cleaned and dried?

a. Once every month	c. At least every six months
b. At least every two months	d. Once per year

Lettering can also use caps. Don't mix capital and lowercase lettering. Additionally, lay out the questions in either a one-column or two-column format. Never mix the two. Be consistent.

MULTIPLE CHOICE TESTS

A multiple choice test is a common method of conducting formal and informal evaluation. In most instances national and state certification and licensing exams are multiple choice, so it's good to administer a number of the tests in multiple exam format to give students ample practice in taking them.

It's important to identify and understand some of the terms associated with multiple choice testing.

An *item* is an instructional design term for all the components of a written examination question, including the question itself and the correct (or best) and incorrect answers.

The *stem* is the part of the item first offered. Often called the *question,* the stem may be written as a question or an incomplete statement.

A *distracter* is a false or incorrect answer to a test question that is designed to be a plausible alternative to the correct answer.

The *key* is the correct (or best) answer to the question.

Considerations Regarding Multiple Choice Questions

An instructor needs to take into account a number of considerations when designing and administering a multiple choice test. The first is bias cueing that occurs from leading students to the correct answer by the way the stem is worded or from the grammar choices. Be cautious on how the answers are worded. Negatively worded stems should be avoided. Try to word the stems in the positive sense. Students are accustomed to looking for positive-worded stems and can be tricked by or may misread negative-worded stems.

Don't have questions build on a previous question's information. The exception to this is in testing the sequencing of steps. It is acceptable to give a scenario and have multiple questions for an independently stated situation.

Avoid questions that pose a fill-in-the-blank segment in the middle of the stem. These are difficult to read, and the meaning may be skewed.

An example of a wrong way to write a fill-in-the-blank question:

1. NFPA standard _______ regulates low-expansion foam.
   a. 10
   b. 1971
   c. 11
   d. 1940

The correct way to write a fill-in-the-blank question:

1. The NFPA standard that regulates low-expansion foam is _______.
   a. 10
   b. 1971
   c. 11
   d. 1940

Avoid using "all of the above" or "none of the above" as a distracter. As soon as the student recognizes one of the other distracters as a correct answer, it eliminates "all of the above" as a possibility. Actually, some instructors may use "all of the above" in an unconscious or misguided effort to extend their teaching into the test. With this approach, "all of the above" is the answer to almost every item that contains this answer, and the students soon figure this out. In the same respect, recognition of a couple of distracters as correct (or possibly correct) leads the student to guess that "all of the above" is the correct answer. In contrast, "none of the above" can be an alternative if the question is one requiring calculation.

TRUE/FALSE TESTS

A true/false test includes a complete statement and a two-choice option of either true or false. This is the most simplistic of test questions.

Considerations Regarding True/False Questions

Using true/false questions limits the answer options to just the two choices—which does not allow for any gray area. It's also difficult to construct good items in a positive voice, yet it is good to avoid negatively worded statements using constructions such as "is not." Also avoid extreme answers that include the absolute statements "always" or "never."

Example of a true/false question:

True or False Firefighters may remove their face pieces to share oxygen with a downed firefighter whose SCBA unit is no longer working.

If using a true/false question for a study guide or even for tests, have the student write what makes the statement false. This makes the true/false question a little more challenging and beneficial for testing the student's cognitive ability.

MATCHING

This type of test question typically offers two columns of information. The student selects items from one column and matches them to items in the other column to form correct or complete statements.

Considerations Regarding Matching Questions

Matching questions work best with definitions and terms or with simple concepts with obvious relationships. They are difficult to properly design, and the instructor needs to be cautious that multiple matches aren't possible. The items used must bear some similarity, and directions of how to match the columns must be as clear as possible. Make it clear to the test takers how often each item can be used. Also, there may be more than one answer per statement or item.

This is an example of a matching question with only one answer and all answers used only once.

Match the ladder to the appropriate description.

___ 1. Nonadjustable in length and consists of only one section.

___ 2. Single ladders equipped at the tip with folding hooks.

___ 3. Single ladders that have hinged rungs, allowing them to be folded so that one beam rests against the other.

___ 4. Adjustable in length and consists of a base or bed section and one or more fly sections.

___ 5. Designed for use as a self-supported stepladder (A-frame) and as a single or extension ladder.

___ 6. Single-beam ladder with rungs projecting from both sides.

A. Extension

B. Roof

C. Pompier (scaling)

D. Single (wall)

E. Combination

F. Folding

COMPLETION TEST QUESTIONS

These are fill-in-the-blank questions. They are designed by using statements with part of their information omitted so that the student must complete the statement.

Considerations Regarding Completion Test Questions

Enough information must be included for students to understand the intent of the statement without being led to the answer. This type of question tends to have answers with an ambiguous meaning, and several answers may be correct. Make sure the statement is clear enough for the appropriate answer, or include in the directions that more than one word may be correct and acceptable.

When creating a question, an instructor should be careful about the length of the space for the answer. The length of the blank may indicate the length of the word. An instructor should create the space to match the length, and do this consistently throughout. It's best to have all the spaces the same length. In the same regard, if there are multiple words to be inserted, have the appropriate amount of spaces for each word; otherwise, indicate in the directions that there may be more than one word for each blank. In this situation it's OK to have the blank in the middle of the sentence. Just avoid having the blank at the beginning of the sentence.

Example of a fill-in-the-blank question:

Double male or double female adapters are often used to connect hose when a pumper set up for a _______ lay is used for a _______ lay.

ESSAY QUESTIONS

An essay question may require a short or long answer. An essay question with a short answer requires a bulleted list of responses or several questions to complete. An essay question with a long answer requires students to provide a lengthy prose-style answer.

Considerations Regarding Essay Questions

An essay question may not be effective for measuring the lower levels of learning. In fact, it's difficult to write an essay question for the lower-level domains.

Essay questions are time-consuming and sometimes difficult to grade, since grading can be very subjective. It is recommended that grading be done in a group format. In other words, if question 10 is an essay question, read all the students' answers to question 10 before grading any. Often if instructors read and grade the answer from each student as they go along, they typically grade the first answers much tougher than the last answers. By reading all the responses, then going back and grading them, there is a better assessment of the answers as a whole. This takes a lot more time and effort, but it's is the most fair and methodical way to grade an essay question. (See Figure 7.8.)

Assigning a point value to the various components of the expected response can be difficult. A rubric is a helpful tool for scoring essay questions because it will describe the criteria for each grade level. When using a rubric, provide the format to the student, especially in instances where you assign a project. (See Figure 7.9.)

FIGURE 7.8 ◆ Grading papers can be a time-consuming process for the instructor.

EXAMPLE

To receive an A the student must provide all the correct information and write in complete sentences without committing any spelling errors, and for a B the student must provide 80 percent of the required answer with no more than one to three spelling errors.

Students often write illegibly because of time pressure or adding information at the end of the answer, making the response difficult to follow. The instructor needs to

Writing Mechanics
Exemplary—3 points
The text has no errors in grammar, capitalization, punctuation, and spelling.

Proficient—2 points
The text has a few errors in grammar, capitalization, punctuation, and spelling requiring minor editing and revision.

Partially Proficient—1 point
The text has errors in grammar, capitalization, punctuation, and spelling requiring editing and revision. (4 or more errors)

Incomplete—0 points
The text has many errors in grammar, capitalization, punctuation, and spelling requiring major editing and revision. (more than 6 errors)

FIGURE 7.9 ◆ An example of one component, writing mechanics, of a rubric for a writing assignment take this into consideration. Students whose thought processes don't flow easily in a linear progression have an unfair disadvantage in a timed test.

Some students may include more information than desired in an attempt to be thorough. Be clear in directions of how much detail the student should provide. This can be accomplished by the amount of space provided for each answer. The instructor can also limit the answer to a set number of pages. In the same respect, if students don't understand the question, they may provide a very well thought out but incorrect answer. In this situation it may be best to award partial credit for a well-constructed incorrect answer.

FINAL TESTS

A licensing or certifying examination is a type of final exam. In most instances, these are multiple choice tests. These examinations are for certification or licensing by a regulating body such as a state or the National Professional Qualification Board. Most certifying or licensing boards require successfully completing one or more of the following examinations to qualify:

A written examination (generally multiple choice)
An oral examination
A practical skills examination

Activity

Construct a 25-question test for a particular topic. Use a test blueprint, and incorporate a variety of types of test questions at a minimum of three different cognitive levels.

POST-WRITTEN EXAMINATION QUALITY REVIEW BY STUDENTS

Once the test is completed and graded, return the test to the students so they can see the results. Determine in advance if the students will be allowed to retain the test. One advantage of students keeping the test is having a learning aid for later testing. Another advantage is it provides examples of the instructor's style of question writing. On the other hand, the disadvantage of allowing students to keep the test is that doing so eliminates the ability to use the test again. This may range from the entire set of test questions to only some of the test questions.

Regardless, allow students to review the test and see what questions they missed. This will highlight their areas of weakness for remedial study. It will also identify areas of weakness in the presentation of the material. When students see what other students missed, or when they can challenge questions, answers, or the wording of a question, it promotes a climate of fairness and manages concerns of bias or discrimination. Reviewing the test can be as useful a learning aid for you as it is for your students. Be cautious about becoming defensive when students challenge the questions. In so doing it may be that the question is not worded appropriately, or there may be more than one correct answer. Be open-minded and set the ground rules and parameters ahead of time.

POST-WRITTEN EXAMINATION QUALITY REVIEW BY FACULTY

Before reviewing the results with the students, compile the results, including an accounting of incorrect answers. It's a good idea to do an item analysis on test score results.

ANALYSIS OF TEST QUESTIONS

Test scoring returns a number of statistics for each test analyzed. It's important to understand the statistical results that can be derived from the scores. The following are some of the statistical reports that can be generated to analyze the scores:

Number of students (N): the total number of students who took the test.

Mean: the average number of questions each student answered correctly on each form. To calculate the mean, add all the scores, and divide by the total number of students.

Median: the midpoint of the distribution where half the scores are above and half below.

Mode: the most scores that are the same.

Standard deviation: the measure of how spread out the scores are on either side of the mean. The formula to calculate the standard deviation is:

sd = $\sqrt{\Sigma(X - X)^2 \div N}$ (the square of the sum of the scores minus the mean scores squared divided by the total number of students)

Reliability estimate (KR-20): an estimate of the correlation between the scores on the test and the scores that would be obtained if students were given a similar test on the same material. Satisfactory levels range from 0.5 for tests of less than 15 items to 0.8 and up for longer tests. The method for computing this estimate assumes that essentially a single area of knowledge is being tested.

T-scores: standardized scores on each dimension for each type of test. A score of 50 represents the mean. A difference of 10 from the mean indicates a difference of one standard deviation. Thus, a score of 60 is one standard deviation above the mean while a score of 30 is two standard deviations below the mean. T-scores express students' performance in relation to each other independent of test length or difficulty.

ITEM ANALYSIS

An item analysis is the best method to determine if any test questions were weak or represented unsatisfactory levels of understanding. Some instructors use the method of throwing out the most missed questions. This is an injustice to the test. Instead, an item analysis evaluates each question to determine whether it was a good one.

Good test questions have most or all of the following characteristics:

- Between 30 and 85 percent of examinees should answer correctly. If the item is too hard or too easy, it contributes relatively little toward ranking examinees according to their knowledge.
- Each of the answer choices should attract at least some of the examinees. If a wrong choice is so obvious that no one selects it, student testing time is saved by omitting it on future tests. There is no need to have the same number of choices for all items.
- The correlations between choices and total score should be positive (preferably 0.3 or higher) for the right choice and negative for the wrong choices. This outcome would indicate that better students tended to get the item right. If a positive correlation occurs for the wrong answer, it indicates that better students were misled into selecting the wrong choice. A poor item correlation can also alert the instructor to a possible mistake in filling out the answer key.

The following are the steps to do a simplified item analysis for a classroom test. If using Excel, set the formulas in the spread sheet and it will calculate the results.

1. Score tests by marking all incorrect or omitted items, and determine the raw score for each paper (number of test items less the number of incorrect or omitted items).
2. Place all the papers (the number of papers = N) in descending order of scores (highest score on top to the lowest score on the bottom).
3. Multiply N (total number of students) by 0.27 or 27 percent, and round off to the nearest whole number. This number is n. For example: If $N = 30, n = 8$ (8.1 rounded).
4. The n lowest scores are the "low group" and the n highest scores are the "high group."
5. Determine the proportion in the high group (p_H) who answered each item correctly by dividing the number of correct answers for the item for the high group by n; that is,

$$p_H = \text{number of correct answers} \div n$$

6. Repeat the procedure for the low group to obtain the p_L for each item.
7. To obtain the difficulty index for the item, add p_H and p_L and divide by 2:

$$p = (p_H + p_L)/2$$

This index must be interpreted with the chance level of the item in mind ($p = 0.5$ for a two-option item marked by all students may indicate little or no knowledge of the point tested). For items attempted by all students, the p-values expected from chance are $1/k$, where k = the number of options for the item.
8. To obtain the discrimination index for each item (D = how well an item distinguishes between students who understand the content universe of the test and those who do not), subtract p_L from p_H:

$$D = p_H - p_L$$

D-values of 0.4 or more indicate a high level of discrimination, and those below 0.2 are low in discrimination.

Here are some of the uses of the item analysis.
If the upper one-third of the group missed a specific item, determine the following:

1. Is the test item keyed correctly?
2. Is the test item constructed properly?
3. Is it free from bias, confusion, and errors in grammar and spelling, etc.?
4. Was the content covered in class?
5. If not, were the students directed to it through self-study?

If the lower one-third of group missed a specific item:

1. Which distracters were most attractive? (That is, which were most often selected?)
Improve distracters that were not attractive.
Consider a distracter well written if it is not selected by the upper third of the class, but it is selected by the lower third.

Activity

Conduct an item analysis using a set of test scores (at least 25 test scores preferably). As part of the exercise, analyze each question, and state whether it was a good discriminator.

FIGURE 7.10 ◆ Oral boards are common for exit exams of many programs. They are also common for promotions in the fire service.

ORAL EXAMINATIONS

Oral exams require verbal answers by a student to questions asked by an instructor or group of instructors. (See Figure 7.10.) The advantages of oral exams are that they evaluate the quick thinking or reactions of the student, giving the instructor(s) the opportunity to assess the student's thought processes.

At the same time oral exams pose a number of disadvantages. They limit the number of students that may be examined at any one time. They are difficult to standardize. The examiner may unintentionally give clues to the examinee in these settings. An oral exam is time-consuming and labor-intensive to prepare and conduct. These exams typically are very subjective so to fairly administer an oral exam, a great deal of attention and concentration is required on the part of both the evaluator and the student. Depending on where the exam is conducted, unexpected distractions can impact the test results. The instructors may be required to evaluate a large number of candidates with little opportunity for breaks, leading to uneven evaluations over time. It may also lead to identification of themes or trends with unfair emphasis or focus on those repeated mistakes—that is, holding successive students accountable for preceding student's performances.

PROJECT ASSIGNMENTS

The advantages to project assignments are many. They are a good tool to get students to work outside the class whether as a group or solo project. (See Figure 7.11.) Project assignments can also evaluate a student's ability to synthesize data. Plus they cater to the specific learning styles or preferences of the individual student. Allowing students to

FIGURE 7.11 ◆ Group presentations are a means for students to work together on a project and then present it to the rest of the class.

select from a group of recommended projects or encouraging them to develop their own helps to promote autonomy and independent learning. If students work in groups, project assignments can help the students develop people skills and conflict resolution skills—both important with fire fighting. For those students who are already working or volunteering as a firefighter, a group project may help hone the art of working together. Students can learn from each other, and stronger students may tutor weaker ones.

Likewise, there are disadvantages to project assignments. They are difficult to standardize. There is also a possibility of plagiarism. If the project assignment is not carefully designed, it can measure only the end product excluding the process; and sometimes the process used is just as important as the final product. For example, learning how to find resources to use in solving a problem or developing critical thinking skills is as important as the project itself.

If the group is required to present the project, the grade should include the content of the presentation as well as the presentation itself. Consider, too, when grading how much each member of the group contributed to the development of the project. If members are not working together or pulling their share of the work, it places unfair workloads on the contributing members. One way to resolve this problem—particularly in light of the fact that the instructor may not have any way of determining just who did what—is to have students grade each other in their group on their contributions to the project. Students tend to be harder on each other than the instructor is on them. Take each grade into consideration—and make sure no one is out to get one of their colleagues.

OBSERVATIONAL REPORTS

An observational report is another means of evaluating students. The advantage of using observational reports is that they can be used for psychomotor or affective evaluation. Observational reports are inherently reliable due to repeated observation. Reliability can be increased by increasing the number of observations.

In contrast observational reports present several disadvantages. The presence of an evaluator may influence student performance, since a student typically performs better when being directly observed. Conversely, an instructor/evaluator may misdirect a student, resulting in the need for retraining. Observational reports are also time-consuming and labor-intensive. They are frequently a one-on-one experience. Developing the criteria for them can be a complex task. Experiences may not be available at the time they are required; for example, if a student is assigned to a station, and the instructor is there observing his/her performance, there may be no calls for the student to respond to with which to observe his/her behavior. This type of observation typically works well by assigning the student with a mentor who will then be responsible for providing an observational report of the student in the real environment.

PRACTICAL EXAMINATIONS

Two basic types of practical examinations are situational and rote. A situational practical exam is a demonstration of a skill in the context of a scenario allowing for manipulation of the outcome or procedure by the student. This is good for evaluating the critical thinking skills as well as skills performance of the student. A rote practical exam is the demonstration of the steps of performing a skill independent of manipulation of outcomes. This generally follows a very specific order of steps.

Practical exams most closely approximate actual job performance and that makes for excellent observation and evaluation of related behaviors and attitudes. In addition, they allow the evaluation of psychomotor skills, decision-making abilities, and leadership skills. (See Figure 7.12.)

Of course, there are disadvantages to practical exams. They are difficult to standardize. A practical exam is time-consuming and labor-intensive to prepare and deliver. There are limited numbers of students that can be examined at one time. The instructor providing feedback needs to be clear about expected outcome, whether a situational or rote response is required and should evaluate accordingly.

PRACTICAL SKILLS EVALUATION

Practical skills are rote mechanical skills. They require simple task analysis by the student. This can be the easiest skill examination to administer, though practical skill evaluations may or may not reflect a student's actual field performance capabilities. Essentially, they are isolated skills performed without "real world" stresses and may not adequately evaluate affective and psychomotor domains.

Testing situational skills evaluates a student's judgment and/or decision making—though it does require more elaborate simulations that are more difficult to develop and deliver. A more accurate predictor of field performance, situational skills tests ask the student to critically think through a scenario that does not always have an obvious answer or just one way to accomplish the same task. This needs to be considered if these are used to evaluate or grade a student.

FIGURE 7.12 ◆ Practical exams are a good method to test the skills the student has learned. *(Photo courtesy of James Clarke, Estero Fire Rescue)*

SIMPLE SKILL EVALUATION

Begin by defining the skill that is being evaluated and determine the degree of expected proficiency of the skill. It is important to select a representative sampling if all the skills in a given area are not evaluated. Make sure the student has a good grasp of the basic concept. Create a written task analysis of the skill if one does not already exist. There are scores of skill sheets already developed. It may mean asking around to find one that is already established.

If a checklist needs to be developed, make it commensurate with the task analysis. Each step should contain some measurable criteria so that all evaluators can agree on criteria of successful completion. Look for established standards, like the NFPA standard for that particular skill, for guidance.

Keep the number of steps to a minimum to reduce errors in evaluation. Allow the evaluator to observe the task as it is performed and complete the evaluation form afterward.

PERFORMANCE EVALUATIONS

Performance evaluations determine and define a student's expected outcome. Two questions will help in the assessment when conducting a student's performance evaluation. First, is skill performance or the decision-making process more important in the situation being evaluated? Second, how stressful or complicated a situation is the student prepared to handle?

Determine what standards will be used to evaluate the performance. The situation should be representative of the desired outcome: realistic environment, realistic scenarios, and obstacles that would occur in the real world. The resources needed for the testing should be evaluated. The higher the domain level, the more realistic the scenario should be. Simulate the situation and responses as accurately as possible. This can be difficult to do depending on the scenario. The instructor needs to keep the situation in perspective and remember the legal ramifications and safety issues surrounding simulated and training events.

List all activities that should be completed in the situation. The activities should be prioritized and listed in start-to-finish sequence. When establishing the grading criteria, weigh the most important aspects and critical criteria appropriately. A checklist is typically the best tool to use for performance evaluation. The checklist should contain:

- The minimum number of properly ordered steps necessary to complete the task
- Steps that are independently observable and measurable
- An outcome consensus understood by each evaluator

During the examination, the evaluator should be free to observe the activity and quantify the behavior and should not be focused on measuring *how much* was performed at each step check. Don't focus on whether the skill was performed or not.

There are instances where the performance is situation-oriented. In these cases what the student does depends on what occurs in the next step. These situations require adequate organization to ensure the outcome of a situation-oriented performance evaluation. A skeletal framework needs to be provided for the evaluator to follow.

Characteristics of Skill/Performance Evaluations

Skill/performance evaluations objectively measure the performance of the student by two methods: an instrument or an observer. These measures need to be replicable. Determine this by answering the following questions:

Does the instrument measure similar performances consistently?
From one student to another?
From one class to another?
From one location (situation) to another?

Fair standards need to be established and known by both students and faculty. When practice sessions are conducted, use an instrument similar to the one that will be used for the evaluation.

Realism is always an important component. Situations and scenarios need to be plausible. Any changes in the condition of the situation or scenario based on the intervention of the student need to be realistic. In some instances, external distractions can be included if they are realistic. Stress can also be added to the scenario, similar to the work environment.

RELIABILITY

Reliability (or stability) is the extent to which an exam is consistent in measuring a student's performance. Let's say that a test is administered to the same group of examinees on successive days, and there was no intervening in the skill tested. If the examinee gets the same or close to the same score on both days the test was given, the

FIGURE 7.13 ◆ Checking the test scores for reliability and validity takes a little extra time, but it is worth the effort to know that the test you are administering is doing what it is designed to do.

scores can be said to be highly stable or reliable. The closer the pairs of scores are, the more stable or reliable they are over time. (See Figure 7.13.)

Other factors influence the change in scores besides time. For example, the examinee may be feeling better or worse than usual; and the noise level, temperature, and ventilation in the testing room all have their effect, as well. For multiple choice tests, variation in luck when guessing can yield score differences between tests. Reliability is affected by all such factors, and it usually isn't possible to determine the relative contribution of each one separately.

Four questions should be answered to determine the reliability of the scores of an exam.

- Does it measure a behavior or body of knowledge consistently on different occasions?
- Does the environment influence consistency?
- Do different test administrators influence results?
- Does it discriminate against groups or individuals?

Coefficients must be taken into consideration when discussing reliability. When there are two sets of test scores as discussed, there is a reliability coefficient that is a Pearson product-moment correlation coefficient between two sets of scores. Correlation coefficients range from -1 through 0 to $+1$, but negative values are not meaningful with respect to reliability. A coefficient of 0 means that there is no relationship between the two sets of scores. A coefficient of 1 would occur if all examinees got the same score on both administrations. A reliability coefficient is signified as rxx. The following statements describe the evaluation of this reliability coefficient.

- High reliability is when rxx = 0.90 or higher. In these cases the test is suitable for making a decision about an examinee based on a single test score.
- Good reliability is when rxx = 0.80 to 0.89. In these instances the test is suitable for use in evaluating individual examinees if averaged with a small number of other scores of similar reliability.
- Low to moderate reliability is when rxx = 0.60 to 0.79. The test in this case is suitable for evaluating individuals only if averaged with several other scores of similar reliability.
- Doubtful reliability is when rxx = 0.40 to 0.59. These tests should be used only with caution in the evaluation of individual examinees. May be satisfactory for determination of average score differences between groups.

In most situations, a test is not administered twice to evaluate reliability. However, the responses from a single administration of a test can be used to estimate reliability. These estimates are called internal consistency reliability coefficients. The Kuder-Richardson formula 20 (KR-20) coefficient is one method to compute the scores. Computation of internal consistency reliability coefficients is based on assuming that the test is not timed and that its content is homogeneous. These coefficients overestimate reliability if there is not sufficient time for nearly all examinees to finish. They underestimate reliability if the test content is not homogeneous. For example, suppose a test contains questions covering two distinct areas of a course and that the number of correct answers for a student in one area is not a very good indicator of how well that student will do in the other area. In this case, rather than combine two disparate topics on the same test, it's better to separate the questions into two subtests, generating separate scores and reliability coefficients.

If there are a small number of examinees or test items, reliability estimates are subject to considerable error—fewer than 25 examinees or 10 items skews the reliability estimate considerably. Keeping information grouped together in the test was discussed earlier. It's particularly important in this instance. If it isn't, the instructor may need to pick out the information or have the test questions identified for a certain topic so that the items can be grouped when assessing for reliability. This also becomes more difficult to do.

In classroom testing, the most common cause of low reliability is test questions that are too easy. When all or nearly all the questions are answered correctly by more than 80 percent of the examinees, the resulting scores will be in a narrow range. For example, under such circumstances, most of the scores on a 50-question test will fall between 40 and 50. In such a case a minimal score fluctuation due to extraneous circumstances (like those discussed above) will have a large relative effect on the class standing of the examinee. On the 50-item test just described, the range for A's might be 47–50. Then a student with a headache, say, who would otherwise get an A, may miss one or two extra questions and get a B. At the same time, a B student who has just moderately good luck when guessing may get an A. These kinds of errors give rise to low reliability coefficients. If the test questions are harder, the scores will be more spread out, and reliability will be higher. Assuming that the same numbers of each letter grade will be given, small errors in scores are then less likely to result in different grades. The standard error of measurement (SEM) is related to reliability: The lower the reliability, the larger the SEM. Of course, it is that average or "true" score that should be used to evaluate the examinee. If the assumptions for KR-20 are met, then the odds are about 2 to 1 (or the probability is about two out of three) that the examinee's true score is contained within one SEM (above and below) of his or her actual score on the test. Recognition of how far the true score of examinees might be from their actual scores may suggest some liberality in determining the cut points between letter grades.

Note that in the above discussion, only scores were described as being more or less reliable. Tests cannot be described in this way. A test may yield highly reliable scores under one set of circumstances and scores of low reliability under another. Influencing factors are administration conditions, appropriateness and difficulty of the test for the examinees, and examinee motivation and attitude. When these factors are unfavorable, the scores are likely to be less reliable than otherwise.

CONTENT VALIDITY

Content validity is the extent to which an exam is representative of a defined body of knowledge—how well a test measures the knowledge and skills it was intended to measure per curriculum objectives. Answer the following questions to determine content validity.

- Are the subtests weighted and distributed properly?
- Do they place an overimportance on a single test?
- Is that your intent?
- Does it cover a reasonable sample of the knowledge and skill objectives?
- Is it an accurate predictor of field performance?

RESOURCES FOR EXAMINATIONS

Peer instructors within your organization may be a good source for examinations of all types. Certification and licensing bodies also may be a source of validated instruments. The local jurisdiction or the state agency that governs the fire service may be yet another resource or governing factor in the design of a test. The following are some of the resources for the various test types.

Written Examination Resources

1. NFPA
2. Publishers' test banks
3. Fire textbooks
4. Fire textbook instructor guides
5. Textbooks of practice certification examinations
6. On-line and computer-based practice certification tests
7. Fire Internet sites

Practical Examination Resources

1. NFPA
2. Fire Internet sites
3. Fire textbooks
4. Fire continuing education programs

Oral Examination Resources

1. NFPA
2. Fire Internet sites

Project Assignments

1. NFPA
2. Fire textbook instructor guides
3. College or university resources
4. Learning styles/preferences information with practical application suggestions

◆ SUMMARY

This chapter demonstrated the need for a mechanism to evaluate the student and identify whether the student is achieving the objectives and goals of instruction. Remember, that testing is not only a measurement of the student's progress but an indicator to make sure the instructor is covering the material effectively. The evaluation and testing process helps determine strengths and weaknesses of the students as well as the program.

This chapter has discussed the program evaluations and how they help the instructor improve the quality of instruction. A new instructor may not design and develop test questions, but should at least have familiarity with the concepts of evaluation and how to construct solid tests. This chapter showed how the instructor should learn how to determine if test questions are well written and match the objectives of the lesson plan and standard. It is essential that instructors construct a valid and reliable test and evaluation process for their classes.

Review Questions

1. List the four levels of Kirkpatrick's evaluation model, and give an example of how to apply them to a fire service class.
2. What are the two types of evaluation? Describe both types, and describe how to use both types.
3. List the five types of examinations. Give at least one limitation for each type of test.
4. Describe how to construct a test blueprint.
5. Give examples of a low-level and a high-level cognitive test, a low-level and high-level psychomotor test, and a low-level and high-level affective test.
6. Describe what an item analysis is and how to do an item analysis on a test.
7. Define and describe reliability for test scores.
8. Describe how to calculate the following:
   Mean
   Mode
   Median
   T-score
   Standard deviation
9. Give an example of at least three test resources.

Bibliography

American Psychological Association. (n.d.). *Standards for education and psychological tests* (revised ed.). Washington, DC: American Psychological Association.

Essentials of firefighting study guide. 1998. Stillwater, OK: International Fire Service Training Association.

Johnson, D. W., and Johnson, R. T. 1996. *Meaningful and manageable assessment through cooperative learning*. Edina, MN: Interactive Book Company.

Judd, R. L. 1998. "The pedagogue's column: The matter of advising on test construction." *Domain3,* Winter.

National Highway Safety Transportation Administration. August 2002. *National guidelines for educating EMS instructors*. Washington, DC.

Popham, J. 1975. *Educational evaluation*. Englewood Cliffs, NJ: Prentice Hall.

Test scoring, retrieved February 14, 2004, from *http://www.testscoring.vt.edu/tests.html.*

Thiel, A., Stern, J., Kimball, J., and Hankin, N. 2003. *Trends and hazards in firefighter training*. (No. USFA-TR-100.) Emmitsburg, MD: Federal Emergency Management Agency.

Waagen, A. 1997. Essentials for evaluation. *ASTD, 9705*.

Winfrey, E. C. 1999. Kirkpatrick: four levels of evaluation. In B. Hoffman (Ed.), *Encyclopedia of Educational Technology*. Retrieved November 17, 2004, from *http://coe.sdsu.edu/eet/Articles/k4levels/start.htm.*

Instructional Media

Terminal Objective

The participant will be able to identify and use various instructional media in a classroom environment effectively.

Enabling Objectives

- Select the appropriate media to use in the classroom.
- Identify and describe the three purposes of media.
- Identify the various types of media to use in the classroom.
- Describe the various types of media used in the classroom.
- Describe simulation and how it can be used in the instructional setting.

JPR NFPA 1041—Instructor I

4-2.2 Assemble course materials given a specific topic so that the lesson plan, all materials, resources, and equipment needed to deliver the lesson are obtained.

4-3.2 Review instructional materials, given the materials for a specific topic, target audience, and learning environment, so that elements of the lesson plan, learning environment, and resources that need adaptation are identified.

4-4.2 Organize the classroom, laboratory, or outdoor learning environment, given a facility and an assignment, so that lighting, distractions, climate control or weather, noise control, seating, audiovisual equipment, teaching aids, and safety, are considered.

4-4.3 Present prepared lessons, given a prepared lesson plan that specifies the presentation method(s), so that the method(s) indicated in the plan are used and the stated objectives or learning outcomes are achieved.

4-4.6 Operate audiovisual equipment, and demonstration devices, given a learning environment and equipment, so that the equipment functions properly.

4-4.7 Utilize audiovisual materials, given prepared topical media and equipment, so that the intended objectives are clearly presented, transitions between media and other parts of the presentation are smooth, and media is returned to storage.

JPR NFPA 1041—Instructor II

5-3.3 Modify an existing lesson plan, given a topic, audience characteristics, and a lesson plan, so that the job performance requirements for the topic are achieved and the plan includes learning objectives, a lesson outline, course materials, instructional aids, and an evaluation plan.

5-4.2 Conduct a class using a lesson plan that the instructor has prepared and that involves the utilization of multiple teaching methods and techniques, given a topic and a target audience, so that the lesson objectives are achieved.

JPR NFPA 1041—Instructor III

6-2.6 Write equipment purchasing specifications, given curriculum information, training goals, and agency guidelines, so that the equipment is appropriate and supports the curriculum.

◆ INTRODUCTION

Technology has made our lives easier. Some will argue that it occasionally makes their life more complicated. Regardless of which end of the spectrum you view technology from, it has become part of the classroom environment. (See Figure 8.1.) Still, the more traditional nontechnology audiovisual aids continue to offer great value in the classroom. It is an instructor's responsibility to determine the type of instructional media that's most important to delivering the message. The instructor is still the best resource for course delivery, and technology is a supplement to instruction. The difference between a good instructor and an excellent instructor is that the excellent instructor doesn't depend on the use of media but uses it to enhance the delivery of the instruction. Let's look at the various instructional media that can be used in the classroom.

◆ AUDIOVISUAL (AV) EQUIPMENT

An instructor has a variety of media and AV resources available for use in the classroom. It will depend on the classroom as to what type of media or AV resources will be accessible. Regardless of what is available, preview all media and test all AV equipment before using them in class. This will also let instructors learn how the devices work so they won't be learning how to use them when they present the material in class. Always be prepared. Instructors don't want to see a PowerPoint presentation for the first time when they show it to the students. If the instructor has difficulty operating

Figure 8.1 ◆ There are a number of media devices available to use in today's classroom.

the equipment, the students will get the impression that the instructor is not prepared for the class; and their confidence level will decrease.

Technology is a great resource to enhance a classroom—but technology has been known to fail at the worse times. Aside from equipment failure, if instructors are unfamiliar with a classroom, they may arrive to find out that the room won't accommodate the resources they have planned to use. A well-prepared instructor has a backup plan. Always have a plan B; in some instances, even resort to plan C.

Trainer's Toolbox

Tips on Using Visual Aids

As an instructor there are a number of issues to consider when using media, such as:

- Watch the lighting in the room.
- Be aware of your movement in front of the screen when projecting media.
- Don't stand between the visual aid and the audience.
- Talk to the audience, not the visual aid.
- Use a pointer whenever possible.
- Be organized and know the material.
- Do not read from the slide for the first time when presenting the material.
- Use the visual aids as an outline; don't read from them.

- Make sure the visual aid is clear and not busy and use a typeface that's easy to read.
- Follow good design for creating PowerPoint slides or transparencies.
- Keep your writing style simple.
- Highlight important items with color, boldface, or italics.
- Check spelling and punctuation; errors are not acceptable.
- Check the equipment prior to the start of the class and have a backup plan if the device doesn't function properly.

SELECTING THE APPROPRIATE MEDIA

Instructional media are used to help students understand the message. They facilitate communication by making the content more easily understood. The word *media* is derived from a Latin word meaning "between" and refers to anything that carries information between a source and a receiver—in this case between the instructor and the student. Differentiated between the medium that carries the message, the message itself, and the methods used. The message is the course content. The methods are the context in which the message is communicated or those processes that the instructor selects to help learners achieve the course objectives.

The use of multimedia in the classroom enhances the delivery of the material only if used properly. Whatever the medium, make sure it is effective for the classroom environment.

PURPOSES OF MEDIA

An instructor should use media in the classrooms to emphasize, organize, and clarify. Let's look at these three purposes.

Emphasize

When using media with text, ensure that visual cues guide students to important concepts and make essential components stand out. The instructor can accomplish this through the use of headings, wide left margins with limited text, boldfaced headings, italics, and a larger font.

Emphasize important ideas; don't force the reader to search for key points. Refrain from using capital letters to highlight because they are difficult to read—though they can be used effectively in short headings.

Trainer Toolbox

Underlining text is not recommended. It has little impact on retention and interferes with the student's own processing and categorizing of the information.

Organize

Use visuals to present material in an organized manner to students. Provide diagrams and flowcharts of sequential steps if appropriate. Research indicates that adult learners benefit from an exercise if they are given printed material and then asked to generate their own "graphic organizer." This could be a simple outline, labeled clusters of circles, flowcharts, graphs, and so on.

Effective organization of material is essential to learning. Material should be reviewed to determine if there is a clear focus to each section. In terms of writing style, start all sections with an introduction and all paragraphs with a topic sentence. Label text so that readers can locate the information they need.

Clarify

Clarity is the cornerstone of understanding. The media selected should clarify difficult concepts. Symbolic representations of concepts, such as words, are not as effective at getting the point across as the real thing.

Use the following rules for your instructional aids:

- Keep your writing style simple.
- Provide ample white space.
- Highlight important ideas with color, boldface, or italics.
- Use a font that is easy to read.
- Eliminate hyphens.
- When using technical terms include definitions.
- Always spell out acronyms on first use and again for new sections.

A variety of media devices and tools can deliver the message in the classroom. The following section discusses the various tools that are available to prepare and deliver media presentations.

OVERHEAD PROJECTORS

Overhead projectors are still available in many classrooms. It's a good idea to have overhead transparencies when the main mode of delivery is computer projection. This ensures a backup plan should the computer fail. The instructor can print overhead transparencies from the PowerPoint file.

Trainer's Toolbox

Overhead Projector

- Use the same frame for all your transparencies.
- Tape a guide on the projector platform so that each image projects onto the same screen area.
- Turn the projector off after discussing the transparency so that the participants focus on what you are saying.
- Be sure the transparencies are in order and none are missing before the class begins.
- Storing transparencies in a three-ring binder in a clear slipcover makes it easier to keep the transparencies in order.
- Do not read from the transparency; instead, give a meaningful and spontaneous lecture associated with the image.
- Using a sheet of paper, you can reveal one section at a time versus showing the entire transparency.
- When presenting a complex idea, break the idea into stages, and use multiple transparencies to convey the message.
- Use pointers directly on the transparencies. This makes for a more clearly concise emphasis of the aspect you wish to highlight.

Before the start of class, ensure that the projector is working and that the visual effectiveness of the overheads is adequate. The lights may need to be dimmed or turned off for the best clarity. Also, any outside light source may affect the visibility of the transparency.

Effective use of the overhead projector is also important. Consider turning the projector off when not in use or have a transparency as a background while discussing the topic. A wise alternative is turning off the projector at various points during the presentation. And a wise investment is having a spare bulb.

Trainer's Toolbox

Tips for Designing Transparencies and PowerPoint Slides

- Horizontal formats project best.
- Pictures are worth a thousand words, so use them—and diagrams, charts, and graphs—whenever possible.
- Keep the message to a simple context, and avoid cluttered designs.
- Rule of thumb for using text is six words per line and six lines per transparency.
- Use key words to emphasize points and remember key concepts.
- Letters need to be a minimum of 3/16″ high. If you don't have access to an overhead projector to check the readability, lay the transparency on the floor with a white piece of paper behind it. If you can read it from a standing position, the audience should be able to read it when projected.

COMPUTER PROJECTION USING POWERPOINT

PowerPoint presentations have become the norm for graphic presentations. A PowerPoint presentation can make or break a class. When PowerPoint originated for use in the classroom, the various transitions and sound effects were a novelty. Words flew in from all directions, noises exploded from the speakers, or animations danced across the screen. These may all be great to add emphasis to a particular point, but too much of anything can be a bad thing. When an instructor overemphasizes animation, flying words, and sounds, the student loses interest in the material being presented and waits to see what will happen next.

In order to use PowerPoint, a computer and a liquid crystal display (LCD) projector are needed. LCD projectors are discussed later in this chapter. Be familiar with how to use the media effectively—and have a backup presentation in case of computer failure. Overhead transparencies can be converted or printed from PowerPoint slides and can be used as backup.

Instructors can build a powerful PowerPoint presentation if they understand its many features. A PowerPoint presentation can be constructed very poorly and actually may deter from the instruction. Note that PowerPoint is becoming a controversial tool in the classroom. Pay close attention to the discussions around this controversy, though PowerPoint will no doubt continue to be used as a resource in the classroom. The bottom line is to use PowerPoint effectively with moderation and don't become reliant on it.

A number of tutorials can help an instructor understand and become proficient in the use of PowerPoint. These can be found in the PowerPoint program and also on the

Internet. Let's look at some of the features of PowerPoint that could be used in constructing a PowerPoint presentation:

- Use Slide Master.
- Record narration.
- Add action buttons.
- Draw AutoShapes.
- Add clip art.
- Insert pictures.
- Create graphs.
- Create organizational charts.
- Create a slide show with Timings.
- Insert a movie.
- Insert digital sounds.
- Add animation.
- Add transitions.
- Add builds.
- Recoloring clip art.
- Display on the Web.
- Add special text effects with WordArt.
- Hide slides.
- Insert Microsoft Word table.
- Print slides and handouts with footers, etc.
- Add hyperlinks and URLs.

Let's talk about a few of the common areas of interest for constructing a PowerPoint presentation.

Text

In most instructional materials, text conveys a message. To optimize the text included in a product, consider what makes good reading. High contrast between background and print enhances readability. **Don't overuse anything!** Just because you can use a particular feature does not mean that you should! Text should be in lower- and uppercase, as appropriate, which is easier to read than just uppercase. Chunking information in clusters of five to seven concepts—which is about the limit of a person's short-term memory—makes the information easier to understand.

Sound

Sounds can add great effects to your presentation. A computer can play CD-Audio using the computer's CD-ROM drive and speakers. An instructor can also record digital sounds to add to presentations. If downloading sounds from the Internet, check on the copyrights. Just because it's on the Internet doesn't give you permission to use it. Non-copyrighted sound files for projects can be downloaded to enhance the presentation—but make sure that the sounds added are realistic and don't create a distraction. Avoid repetitive sounds even if they are funny the first time.

The following extensions are a list of some available sound file types.

- .au: Audio, UNIX, and Internet
- .aiff: Audio and Internet—Apple and Macintosh Media Players
- .mov: QuickTime media file
- .wav: Wave, Windows and Web

- .voc: Sound Blaster
- .midi: Musical Instrument Digital Interface, synthesizer and Web
- .mpeg., mpg: Internet
- .mp3: MPEG-1 Layer 3 compression, high-quality Web
- .rm, .ram: Real Player
- .wma: Windows Media Player

Creating Sounds

Creating sounds is a fairly easy process. A recording device is needed, of course, and a microphone. The sampling rate and the length of the sound will determine its file size. The sampling rate is the number of samples per time expressed in kHz. For example, telephone sound is sampled at 4 kHz, FM radio at 16 kHz, CD audio at 22 kHz. As a rule of thumb, each kilohertz takes one kilobyte per second, so 5 kHz takes 5 Kbytes per second, and 22 kHz takes 22 Kbytes per sec. Once a sound is captured, named, and saved, it can be edited using a commercially available sound program. Sound Forge is an example of a program for this task. Sound files may be inserted in PowerPoint slides.

Trainer's Toolbox

Sounds Sources

If looking for sound sources on the Internet, try a few of these sources.

- History Channel, for speeches: *http://www.historychannel.com/speeches/index.html*
- Library of Congress, for speeches: *http://memory.loc.gov/ammem/ndlpedu/find.html*
- National Public Radio archives: *http://www.npr.org*
- Broadcasts: *http://www/broadcast.com*
- Multimedia Sound: *http://mmsound.about.com/*

Graphics

Graphics are image files used for visual display in documents, presentations, and Web pages. Graphics have the following file extensions:

- .bmp, .pict: Bitmaps
- .jpeg, .jpg: (Joint Photographic Experts Group)
- .gif: (Graphics Interchange Format) these can also be animated
- .tif: transfer image file format
- .png: Portable Network Graphics
- .pcx: pixel graphics format

Creating Graphics

Graphics can be obtained from the clip art files of many applications or saved from the Web. Saving images from the Web begins with a mouse click. Right-click over the image, choose "Save Image As" from the pop-up menu, and then choose a name and save-to location. Remember to take copyright into consideration when saving images off the Web.

To create images, use a drawing program such as PhotoShop, Paint It, Color it, or Paintbrush. Windows Accessories includes Paint for creating bitmaps. To capture images, peripherals such as scanners or digital cameras are used. An object or printed image can be scanned and saved as an image file. Digital cameras capture digital photographs that are transferred to a computer for editing and saving. These devices are discussed in more detail later in this chapter.

Video and Animation

Digital video and animation can also add realism to presentations. Video can be saved from the Web or created using software and hardware. Video capture cards or digital video cameras will capture digital video from an analog source such as a VCR or TV. Digital video cameras can capture live video, which can be saved as a digital movie file.

Animation is a sequence of still images (usually GIF images). Software such as GIF Construction Set or GIF Animator can be used to combine images into a file.

The various types of video file extensions are:

- .avi: audio/video interleave, Video for Windows, Web
- .mov: QuickTime movie, for applications and Web
- .mpeg-x: compresses each frame up to 50:1, for CD, DVD and broadcast; loss varies, for applications or Web
- .qtw: movie files in the QuickTime for Windows format

Trainer's Tip

A video signal is 10,000 times larger than a voice signal of the same duration. A frame of digitized video can occupy over 750 KB: 1.3 GB for 30 frames per second (fps). Video files can be compressed to reduce their file size that makes storage easier and shorten loading time.

Activity

Create a PowerPoint presentation. The presentation should include various media used effectively: animation, sound, text, transitions, video, and still images. The presentation should last no longer than 10 minutes.

SLIDE PROJECTORS

Slide projectors have not been retired completely to local museums. A number of authors still augment their texts with 35mm slides. A 35mm slide produces a much sharper picture than an overhead projector and doesn't have the keystone effect—when the picture is wide on the top and narrow on the bottom—that is sometimes common when using an LCD projector. Many of the LCD projectors have functions to correct keystone effects.

Still, slides present a number of disadvantages. First, they require more storage space. Second, it's difficult to make text slides unless you get them commercially produced. Third, slides also have a bad habit of spilling prior to your presentation. (If using slides, make sure that each one is numbered so that if the slides fall out of the carousel, they will be easy to put back in the proper order.) Fourth, slides don't give you the latitude to move around in the visual presentation like overhead transparencies. Finally, the room generally has to be dark for them to properly project—and a darkened room is neither conducive for note taking nor for attentive learning.

Trainer's Toolbox

Using Slides Successfully

- Make sure the slides are properly sorted and right side up. Here's how:
  - Arrange them in order, and number them sequentially.
  - Hold the slide as it will be seen on the screen right side up, with the letters running left to right.
  - Place a mark or a number on the bottom left corner. This spot is referred to as the thumb spot because when the slide is placed upside down in the projector, your thumb will be on top of the spot.
- Use words on title frames to cue the participants to upcoming subject information.
- Prior to beginning the class, make sure there is a means to illuminate your notes once the room is darkened.
- Insert a black or gray slide for those periods between slides when you will be lecturing and the screen should be blank.
- Keep the commentary on each slide to less than a minute unless describing a complex visual.
- Incorporating music to create a mood or capture attention is an effective media technique.
- Begin and end with a black or gray slide, not a white screen.
- Use a remote-control device to advance slides. This gives you flexibility to move around the room instead of being essentially tied to the projector.

VIDEOTAPES

A variety of programs utilize videotapes as a supplement. (See Figure 8.2.) As with any other form of media, make sure that everyone in the classroom can view the TV screen. This may mean elevating the television and making sure the screen is large enough: a 27″ to 32″ screen is recommended for most class sizes. Anything less than a 20″ screen is inadequate in any classroom setting. A number of LCD projectors can have a VCR connected to project the video onto the screen.

Sound is another important element of video presentations. Be sure the volume level is adequate for all students in class. It should be neither too loud nor too soft. Lighting, too, is a factor for adequate viewing. Outside light may need to be blocked to prevent glare plus dim the lighting in the room.

Videotapes are typically copyrighted and should never be duplicated for any reason without the appropriate permission. Any supplemental videotapes that are suggested for various programs can be purchased by the instructor for presentation.

DVDs are becoming more prevalent in the instructional arena. DVDs can typically be used in a number of computers, which makes it easier for the instructor to deliver the

FIGURE 8.2 ◆ Video programs are good to use as part of the classroom instruction.

media without a number of different devices. DVDs are also easier to maneuver to the part of the video to project. Since the built-in speakers in laptops don't always project the level of sound necessary for the classroom environment, the instructor can connect external computer speakers to generate the level of sound needed.

Videos and DVDs can be produced rather economically. Students are not interested in watching home movies; however, a number of quality video programs developed by amateur emergency service personnel have sufficient quality to be used in the classroom setting. If creating your own video, start by investing in a good digital video camera. In addition, invest in digital video editing software. A favorite among instructional technologists is Final Cut Pro for Mac or Adobe Premiere for PCs. Both programs are relatively inexpensive, easy to use, and produce quality videos. Digital video takes up a large amount of storage, so the computer will need to have a minimum of 60GB of storage space.

Trainer's Toolbox

Tips for Video and Film Presentations

- Check the lighting, seating, and volume settings to be sure that everyone can see and hear the presentation.
- Preface the presentation by briefly reviewing the objectives of the lesson, listing the main points on the board.
- After the presentation, review the main points.
- A smaller, brighter image is better than a large dim one. Move the projector closer to the screen or cover the windows with paper if necessary.

VISUAL PROJECTORS

Often while presenting information, it would be effective to project a chart or text from a book to the classroom. Visual projectors are an excellent audiovisual tool that can be used to easily project photographs, books, or other media onto a screen. (See Figure 8.3.) The visual projector allows you to zoom into portions of a book or object to allow for a better visual for the participants. This works better in most instances then trying to scan a document and portraying it through computer projection.

The projector looks very similar to an overhead projector, but its capabilities and cost far exceed that of an overhead projector, making this device more practical for training centers or classrooms that are used on a regular basis.

DIGITAL CAMERAS

Digital cameras are now more common then 35mm cameras. They come in many shapes and sizes. It is advisable to do some research before going out to purchase one or a person can be overwhelmed by the choices. One of the most important issues to determine is the number of megapixels needed. Megapixels are the number of dots you have per square inch. Think about it this way: the more dots there are in a square inch, the less grainy the picture will be. So the fewer the megapixels, the lower the cost—but the fewer options there are in photo reproduction. In most instances, don't go lower than a 3.3-megapixel camera.

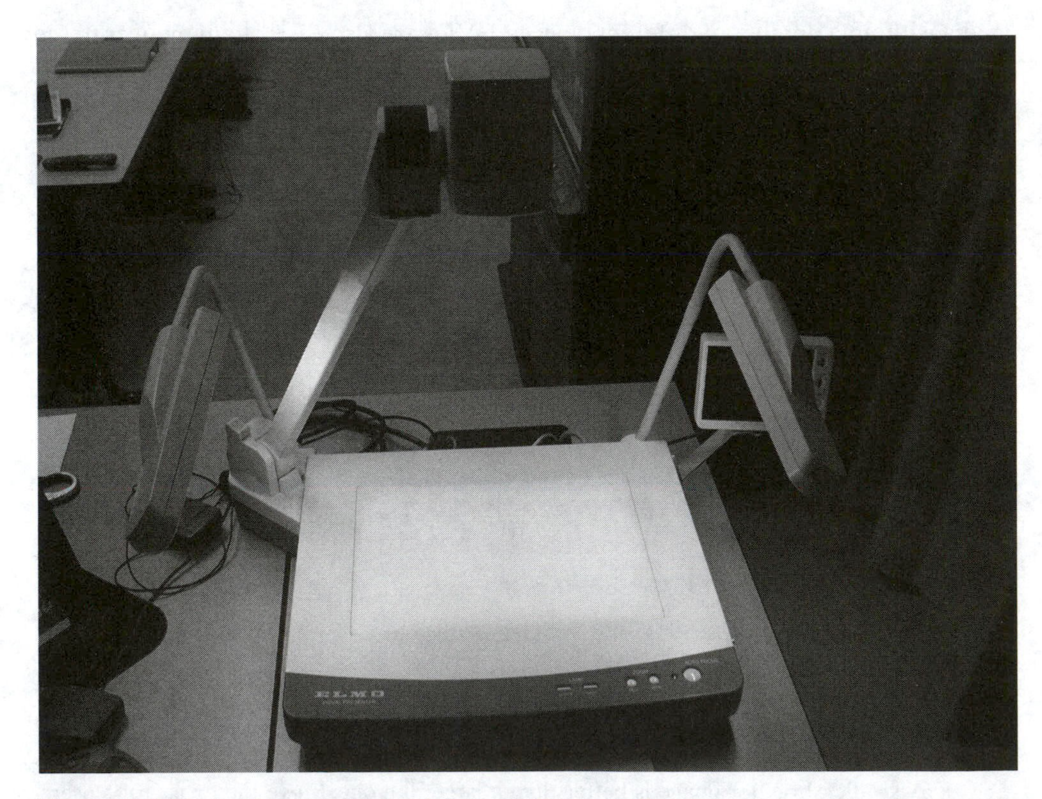

FIGURE 8.3 ◆ A visual projector is an excellent media resource to use to show pages from books or other documents.

When using the camera for print reproduction, the greater number of megapixels the better. But the reverse is true for use with computer projection. Photos need to be small file sizes when using them in computer presentations or uploading them to the Web. This will allow the picture to project or download faster.

Digital cameras are great for taking photos to incorporate into media presentations. (See Figure 8.4.) Copyright is a huge issue in today's world of easy Internet access where it's so easy to find a picture that fits the subject and insert it into the media. A digital camera can be used to take photos without the worry of copyright ownership. (Refer to Chapter 5 for further information on copyright information.) Caution needs to be taken when photographing individuals. If they are in a public location, typically it's acceptable to photograph them. However, if the person is in an environment that is not public domain, it is not acceptable to take a picture. An example of this would be taking a photo of a patient in an ambulance, which is not a public setting. If the person's signature is obtained on a release form, the photo can be used.

Digital photos can have a number of other benefits for instructors. Images can be captured of local buildings and scenery to make a presentation more adaptable to the environment in which the students typically function. This can include the equipment and trucks they are accustomed to using. In addition, using a photo editor like Adobe Photoshop, the instructor can take photos of various buildings within the response area and create a scenario with smoke showing from certain portions of the structure.

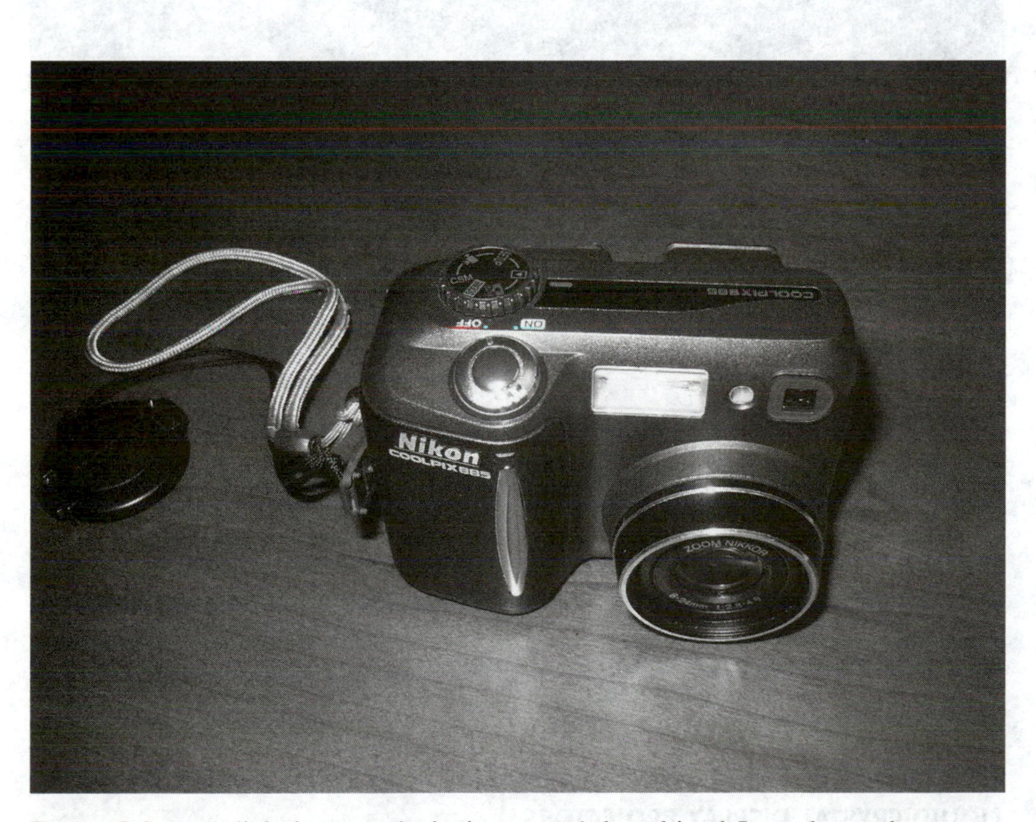

FIGURE 8.4 ◆ A digital camera is the instructor's best friend. It can be used to capture images to use in later classes.

FIGURE 8.5 ◆ A scanner, copier, and printer can all be found in one device, which saves space in the office.

Digital cameras are a great tool for instructors and with their relatively low cost they are a tool no instructor should be without. A number of digital cameras can also capture short segments of digital video, but usually they are unable to capture video of any true length. (See the earlier section on digital video cameras.)

SCANNERS

Scanners are a great accessory to add to a computer system. (See Figure 8.5.) A scanner allows the user to scan documents or photographs into a digital image to upload into the computer. Just as with the digital camera, when selecting a scanner pay special attention to the resolution. If the purpose of scanning a photo is to place the photo in printed material, scan the image at a high resolution. When scanning a text document, an optical character recognition (OCR) software program is needed to convert the image of the text to a true text document for editing. Scanning text tends to be very difficult and requires the user to edit the text in most instances. This can be very time consuming and it may be quicker in some instances to have someone type the document into the computer.

LIQUID CRYSTAL DISPLAY PROJECTORS

A liquid crystal display projector—or LCD projector—is a critical device for any instructor wishing to project media. (See Figures 8.6 and 8.7.) An LCD projector

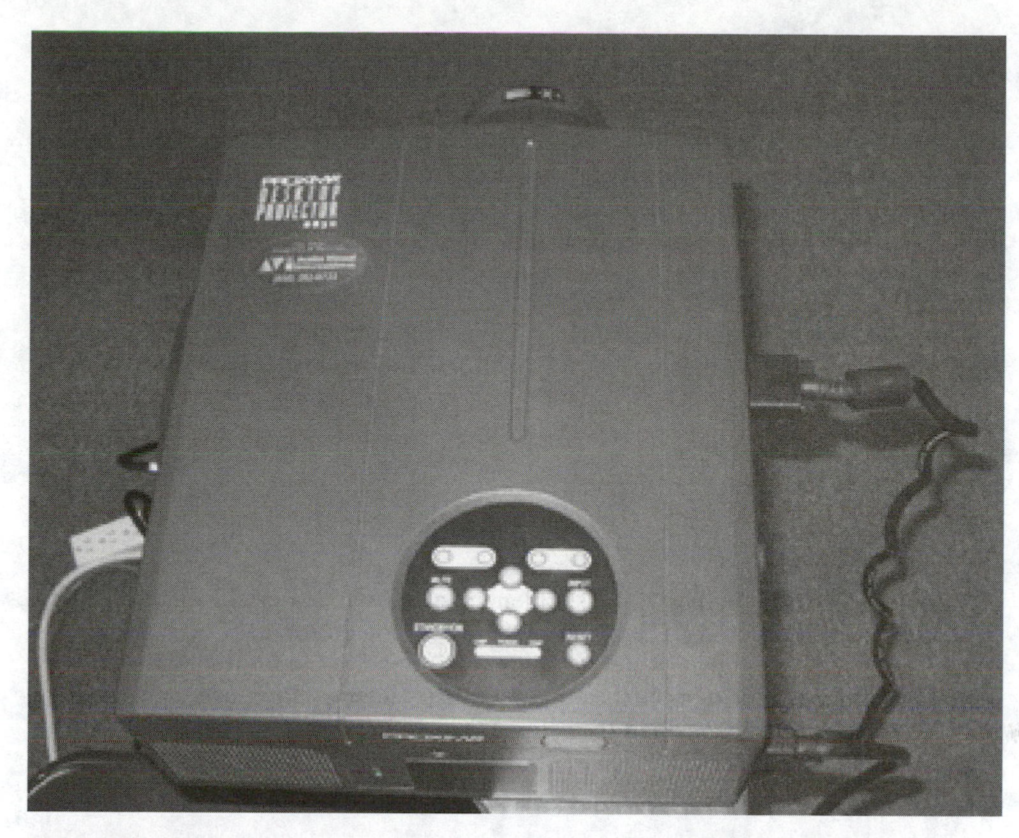

FIGURE 8.6 ◆ LCD projectors come in many sizes, including this large one.

FIGURE 8.7 ◆ Small LCD projectors save space and are easy to transport from location to location.

can be used to project documents and images from a computer and most can be connected to a VCR or DVD player, but it's important to check out the features prior to purchasing an LCD projector. Many newer projectors are much smaller than the ones previously on the market, and some of these can't project video from a VCR or DVD player. The other issue is sound. A number of LCD projectors have the sound system built into the projector; depending on the size of the classroom, the sound may not be able to project sufficiently for everyone to hear. If you are teaching at another facility, and the facility is furnishing the equipment, have a clear understanding of the type of AV equipment you need and what your host will provide.

◆ HANDOUTS

Handouts can be an effective way to supplement classroom material with the latest information. (See Figure 8.8.) These handouts can be copied for classroom use as long as they are not copyrighted. (Refer to Chapter 5 for additional information regarding copyright and the education fair use right.) Most students like receiving something tangible to take with them when the seminar or class session is over. A handout can meet this need and expectation.

FIGURE 8.8 ◆ Handouts are a good supplement to distribute to students as part of the class.

Trainer's Toolbox

Handouts with the instructor's name on them are an excellent method for the student to review the material taught in a class or seminar and for the instructor to gain recognition.

PADS AND EASELS

A large drawing pad can be used to emphasize points or to illustrate a situation that may be difficult to describe. The written information can be used again in future classes.

A pad and easel can be used effectively for participants in the classroom in group settings. (See Figure 8.9.) The pad can be used by each group to write the responses during group projects. The pages can then be posted around the classroom for group presentations or for later reference during the class.

CHALKBOARDS OR ERASABLE BOARDS

Chalkboards are probably almost as old as education itself. Chalkboards, too, can be used to illustrate topics that may be difficult to describe. Today, though, erasable or white boards may be more effective. (See Figure 8.10.) Use a variety of colors on these boards to get the message across. A bonus: Erasable boards don't create dust like chalk,

FIGURE 8.9 ◆ An easel is portable and can be taken from location to location as well as moved around the classroom.

Figure 8.10 ◆ Erasable boards have virtually replaced the chalkboard. The erasable board does not create dust like chalk and helps those with allergies to dust among other benefits. *(Photo courtesy of James Clarke, Estero Fire Rescue)*

which helps those who suffer from dust allergies. Remember to keep the student in mind when using this medium. Appropriate legibility is very important to send the message to the students. Block printing is better than script when writing on these boards.

Activity

Set up each of the AV devices at your institution as though you are preparing for a class session. Go through the process of troubleshooting problems for each piece of equipment. Also, determine the best placement of the devices as they relate to the classroom and lighting.

◆ VIRTUAL REALITY

Virtual reality (VR) is a common instructional strategy used in simulations. A 1995 paper authored by H. Rose discusses seven steps for students to solve problems using VR. They are:

1. VR may prove to be a powerful visualization tool for representing abstract problem situations.
2. Virtual worlds allow for a high degree of trial and error that may encourage students to explore a greater range of possible solutions.

3. The student is free to interact directly with virtual objects that allows for firsthand hypothesis testing.
4. The virtual world can be programmed to offer feedback that focuses the student's attention on specific mistakes, thereby enhancing students' ability to monitor their own progress.
5. The VR system can collect and display complex data in real time that may help students obtain their desired goals.
6. The immersive nature of VR might enhance students' capability to retain and recall information that could facilitate the evaluation of solutions.
7. The virtual world is a fluid environment well suited for the iterative process of refinement. (Rose 1995, 21)

SIMULATIONS

A simulation is an environment that's created to place the user in a position of experiencing a real environment. A simulation can be described as an environment that is created on the external realm. This is a phenomenon where our sensory organs are stimulated to the level that is outside our limits. Basically, what occurs is, the perception in this phenomenon has our thoughts originating externally versus internally.

As noted earlier in the text, adult students learn best when they have hands-on experience that can make the learning practical. Simulations are an excellent means to immerse the student in as realistic an environment as possible. Simulations can be everything from a computer simulation to a mock town. (See Figure 8.11.)

Driving simulators are one example of simulation that may be used in the fire service to teach a skill. Simulators are divided into different subsystems. They include

FIGURE 8.11 ◆ Simulators are becoming an integral part of the classroom. A driving simulator by Road Safety is pictured here. A student is driving a simulated competency course before driving an actual emergency vehicle on the real course.

FIGURE 8.12 ◆ Students like practicing their skills in a simulated environment. In this scenario a firefighter is simulating that he has been overcome by some type of chemical. *(Photo courtesy of James Clarke, Estero Fire Rescue)*

visuals, sound, force feedback, vehicle model, and scenario. When all the systems work together, they create the simulation of driving or operating a vehicle. (See Figure 8.12.)

One critique about simulators is that they don't feel like the real thing. But most simulators have some type of force feedback—the reactions or the "feel" that the driver experiences when operating a vehicle—including braking and accelerating, cornering, suspension and road elevation, suspension and cornering, and steering. The complexity of force feedback depends on the simulator.

Simulators can be traced back to the early 1950s and have a long and rich body of scientific and technical literature about their use for training. Some training environments that are too complex and too difficult to create a prototype—or too dangerous to test in the real setting—are ideal for simulation.

Ross (2002) listed a number of reasons why simulations are different from traditional classroom environments:

1. They can be set up immediately.
2. A wider variety of situations can be replicated compared to any other method of training.
3. Records and results are automatically and objectively gathered and logged.
4. They are easily repeatable, adding a dimension of consistency for benchmarking.
5. They are more accessible to the student and the training department.
6. They are more resource friendly.
7. They are inherently safer than a live exercise.
8. They deliver cost effectiveness.

The Phoenix Fire Department has converted one of their old fire stations to a simulation training center where the fire department can apply the SOPs and

FIGURE 8.13 ◆ The center in Phoenix has a simulated command vehicle. *(Photo courtesy of Don Abbott, Phoenix Fire Department)*

Incident Management System. This state-of-the-art Command Training Center (CTC) was opened in August 2000. The CTC, which is comprised of a large simulator with a classroom, is the basis for Phoenix FD's incident command (IC) certification program. (Figure 8.13.)

Seven main areas are featured at the center. The first area is the command van prop, a fully equipped replica of the vehicle used at the scenes of first alarm and greater incidents. There are eight seats in the van for the IC, support officer, senior advisor, and all the section chiefs.

The second area showcases tabletop models. Don Abbot, who is well known in the fire training industry for his development and training using the Abbotville tabletop training simulator, created a replica of downtown Phoenix. (See Figure 8.14.) The tabletop simulation creates a different perspective for the trainees to view the scenario they are being shown.

The suburban command post prop is the third area of the CTC. (See Figure 8.15.) As in most fire agencies, the IC is typically the first arriving battalion chief (BC). The CTC replicated the vehicle that the BC operates by taking one that was going to be scrapped and cutting the front off the vehicle, then equipping it with the usual radios and other equipment found in a BC vehicle. This prop can either be positioned to overlook the tabletop simulation, or a portable screen can be placed in front of the windshield to project a fire simulation.

The fourth area is the control room in which video feeds, the lighting system, and the computer controls are located. The room houses two facilitators who

Figure 8.14 ◆ Tabletop model of downtown Phoenix *(Photo courtesy of Don Abbott, Phoenix Fire Department)*

Figure 8.15 ◆ Command post for the simulator center *(Photo courtesy of Don Abbott, Phoenix Fire Department)*

Figure 8.16 ◆ The control center houses the computer brains of the center. *(Photo courtesy of Don Abbott, Phoenix Fire Department)*

operate the simulation progression controlling the images and the progression of the scenarios. (See Figure 8.16.)

The CTC design includes the sector kiosks or small cubicles. (See Figure 8.17.) When the Phoenix FD team created the design, they noted that the sector position was not part of most of the commercially developed simulation programs. The CTC includes 12 sector kiosks with each cubicle equipped with a flat-screen monitor, a headset, and a radio. Through the software program the CTC uses for the simulation of fires, the images are projected on the screens of the sector officers. The strategies employed by the officers dictate the progression of the fire scenario. (See Figure 8.18.)

A dispatch center comprises the sixth area of the CTC. It copies the design and emulates the dispatch center found in Phoenix. As is typical, a dispatcher sits in the center and dialogue involves the dispatcher as it does during a real incident.

The seventh and final area of the CTC is the classroom. (See Figure 8.19.) Able to accommodate 25 to 30 students, the classroom is equipped with AV support necessary for teaching. This is where the participants can assemble to prepare for the drill scenario. After the completion of the drill, the participants can then reassemble to discuss and debrief how the exercise actually went. (See Figure 8.20.)

FIGURE 8.17 ◆ There are a number of kiosks that various individuals sit at during an exercise and view the progression of the fire on their screen. Each individual has a headset with communication capability. *(Photo courtesy of Don Abbott, Phoenix Fire Department)*

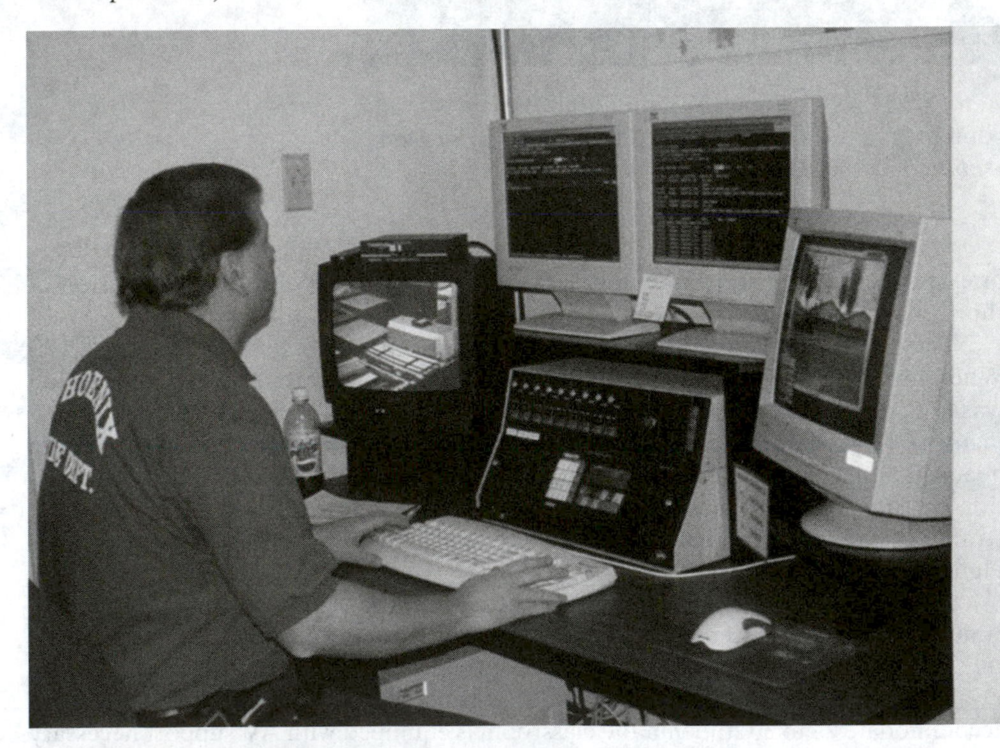

FIGURE 8.18 ◆ A simulated dispatch console is in the center to make the entire exercise complete with simulated dispatch and radio communications. *(Photo courtesy of Don Abbott, Phoenix Fire Department)*

Figure 8.19 ◆ The classroom can be used for orientation and debriefing of the exercises. Plus additional training sessions can be conducted. *(Photo courtesy of Don Abbott, Phoenix Fire Department)*

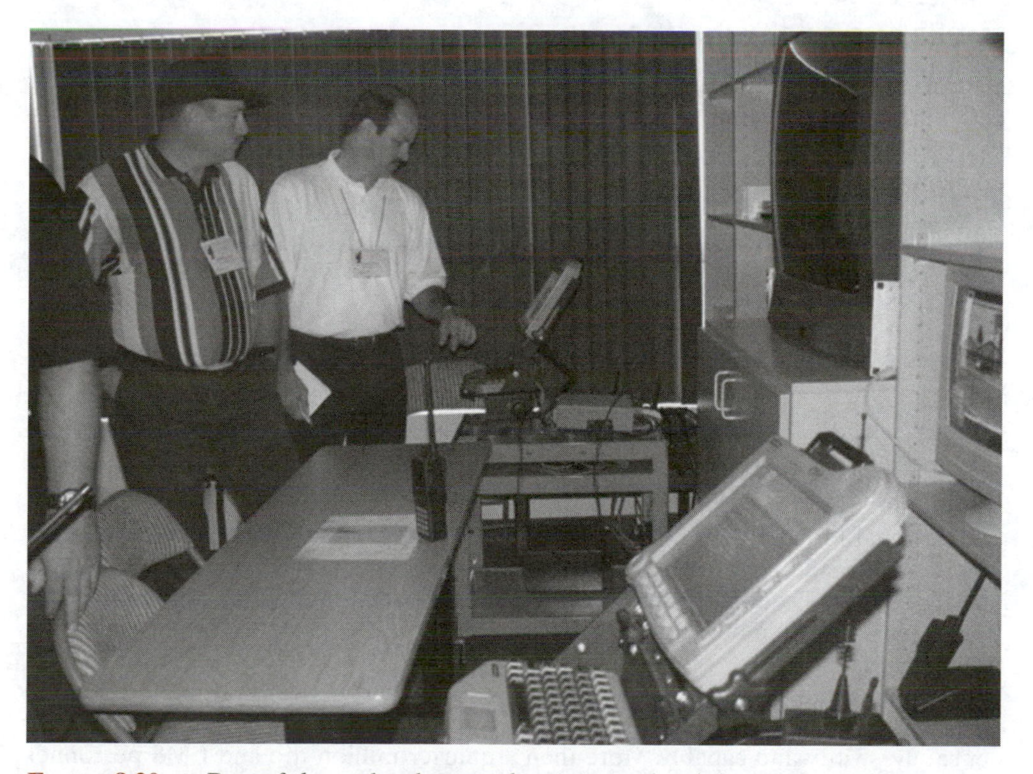

Figure 8.20 ◆ Part of the technology at the command training center *(Photo courtesy of Don Abbott, Phoenix Fire Department)*

FIGURE 8.21 ◆ The Phoenix center has a small-scale simulated city with fire simulated on one of the buildings. *(Photo courtesy of Don Abbott, Phoenix Fire Department)*

SMOKE AND FIRE SIMULATORS

Acquiring structures for conducting live burns (see Chapter 5) is increasingly difficult. Simulation is becoming a standard in today's world. (See Figure 8.21.) A number of simulations are available to emulate nontoxic smoke and simulated fire environments. Simulation is a means to enhance and improve techniques and skills in firefighting—though it's important to note that it is difficult to re-create an environment that's similar to the real thing without the risks associated with a real fire event.

◆ SATELLITE TV

Satellite television has become very prevalent in the world of technology. Many fire stations are already equipped with a satellite dish for television reception from local and national broadcasters. For an additional cost, a person can connect to fire and emergency broadcast stations. The most popular network for the fire service is the Fire and Emergency Training Network (FETN). This particular network requires an additional satellite dish in most cases, but it can be part of the package for your agency's training program.

For more than a decade, FETN has developed and delivered training to emergency service personnel across the country empowering emergency personnel to react safely, swiftly and capably. More than a quarter million fire and EMS personnel watch the programming each month.

What Is FETN?

The training program that FETN develops and airs is compliant with NFPA, DOT, ISO, OSHA, and other standards. The great thing about a program such as FETN is that it brings those who are experts in their field right into your station to train your personnel. This can augment the teaching program and improve the quality of training personnel are receiving.

FETN states that the primary objective is to help the training officer get the job done. FETN also keeps the instructor and the department on top of new technologies and standards, and they're committed in aiding the community to maximize the available ISO (International Organization for Standards) rating points in training. According to FETN, they achieve this objective by:

- Providing ease of access to their programs and services.
- Keeping your service in your district while receiving your training.
- Focusing on depth.
- Providing information on new standards and requirements for the fire and EMS professions.
- Maximizing your points in the training area for ISO (helps lower your community's insurance rate.)
- Providing basic continuing education for fire, EMS, hazmat, and other technical rescue specialties.
- Providing blood-borne pathogen training.
- On-going receipt of federal-level emergency management training.
- Providing fire/EMS–specific news from around the country every business day.
- Providing third-party documentation of your training.
- Providing effective on-going training in the station, thus keeping your firefighters deployed.
- Providing training developed by nationally recognized experts.

How Is FETN Packaged?

FETN has multiple venues for providing training, including satellite, video, Internet/Intranet, or CD-ROM.

Satellite Delivery

This mode of delivery is the most comprehensive program offered by FETN. FETN installs a satellite dish at the facility and delivers 16 hours of training and news daily. There is late-breaking news through their affiliation with CNN, and program guides, lesson plans, and testing materials are sent to the subscriber every month. The FETN satellite package airs the following programming:

American Heat: This programming is incident-based real-life training that meets NFPA standards. *American Heat* gives personal accounts of real incidents that have occurred.

Pulse and *PULSE Plus:* These training programs are designed for both the ALS and BLS provider. As these are CECBEMS-approved programs, the EMS providers in your agency can get CEU's for completing them.

Protecting America: After September 11, 2001, an increased awareness and need for more training in response to terrorism incidents was critical. This programming meets that demand.

Survival: This program gives hands-on exercises designed to enhance firefighter survival and safety. This programming meets NFPA standards and reinforces the "save our own" philosophy and techniques.

Hazmat and *Technical Rescue:* These programs help a department meet the federally mandated training requirements. Additionally, they reinforce standards for awareness, operations, technician, and incident-command training.

Back to the Basics: These programs provide IFSTA-based and hands-on exercises that reinforce the basic concepts for fireground safety and effectiveness.

Daily News: FETN has an affiliation with CNN to bring up-to-the-minute news on fire and EMS-related news stories. An hourly news block keeps personnel current and up to date on the legislative and political news affecting the department and community.

WEB-BASED LEARNING

With the explosion of the Internet over the past few years, many students now expect courses to be offered and available through Web-based learning. The initial fear of instructors was that e-learning would be their demise. Well, to paraphrase Mark Twain, the news of their death has been greatly exaggerated. As a matter of fact, instructors are needed whether the course is being delivered on the Web or in a traditional classroom. (See Figure 8.22.) A number of courses in the fire service need face-to-face delivery and can't be taught via e-learning. On the other hand, Web- or computer-based courses can be very beneficial.

Designing and developing computer or Web-based classes can be difficult. In the beginning, a number of these Web courses were what is called electronic page turners—essentially text on a screen. Web-based or computer-based courses need to

FIGURE 8.22 ◆ Computer training can occur anywhere.

be interactive, which is very challenging for instructors to design and develop. One hour of e-instruction typically takes a minimum of 200 hours to develop—a vast amount of time that most instructors don't have. A number of companies specialize in e-learning. Unless an instructor has a background in designing and developing computer-based courses, coupled with enough time to do the job right, look at outsourcing this component of the training program. The other factor to consider is the cost of developing computer-based courses.

If instructors plan to instruct a Web-based course, the demand on their time becomes greater than a traditional course. In the traditional setting the class meets at a set time, versus a Web-based course where students can access the course at any time. Additionally, students expect instant response from the questions or assignments they send. Guidelines must be established in the beginning of the course as to how fast an instructor will respond to the students with answers.

"Blending learning" is an alternative and may be the best way for an instructor to use Web-based learning. Blending learning is conducted in a traditional classroom setting but includes assignments, activities, and resources on the Web—a blend of traditional classroom and technology in one class. Texts are written specifically on distance learning and Web- and computer-based learning. If this is something of interest, read more about this topic and take specialized courses to learn more about this instructional technique.

◆ AUDIO

Depending on the classroom and the number of participants, a microphone may be needed. Microphones can typically be categorized into two categories: wireless and wired. A wireless mic—which can be attached to a lapel or hand held—offers the most freedom for the instructor to move around the classroom. Wireless microphones may have issues with frequency and distance limiting how far from the remote you can actually move.

Wired microphones can be fixed to a lectern or be handheld to give the speaker more freedom to move about the audience. The wires create a tripping hazard, and the instructor needs to be careful not to get tangled in them. A wired system tends to have fewer problems than a wireless system as far as technical difficulties.

◆ AUDIOCASSETTE

Audiocassettes are very useful for instructing classes on radio communications. An audio recording from a 911 dispatcher taking a call, from the radio communications of the dispatcher to the completion of the call, can be used to critique incidents and to teach students radio communications. Audiocassettes are relatively easy and inexpensive to reproduce.

◆ REVIEW OF SELECTING MEDIA

Adding a variety of media to a presentation helps to keep students interested and maximize various student learning styles and preferences. Media comes in as many price ranges as there are forms—and expensive is not necessarily better.

To determine whether the media that is planned for use is a useful resource, it needs to:

- Be appropriate for the audience
- Be professionally presented
- Be accurate
- Target students' reading and comprehension levels
- Cover an appropriate depth of information
- Contain current information, including trends and updates
- Promote good behavior and practices in students (example: wearing PPE when appropriate)
- Be easy to use

Regardless of which media chosen, it should be defendable and credible. Remember:

1. Select media from a refereed journal or a peer-reviewed Internet site.
2. Don't assume because it was commercially prepared that it's designed well or the content is accurate.

A refereed journal or a peer-reviewed Internet site is essentially information that has been reviewed by a panel before it is published. For example, many of the professional journals in the industry have a panel of reviewers to look at the articles submitted for publication in order to make sure the article is acceptable.

In the same regard, just because something is commercially produced does not make it credible. Be careful and closely scrutinize any media before using it.

SUMMARY

Many media devices can be used to develop or deliver instructional materials. Many were discussed in this chapter. Technology is ever-changing, and it's sometimes impossible to keep up with the latest and the greatest technology. Regardless, such basic media devices as handouts and erasable boards can be just as effective when used in the proper format. Media is designed to emphasize, clarify, and organize the material being delivered. Don't allow it to take away from the message. The latest technology is great, but it can be ineffective if used improperly.

Review Questions

1. Describe the three purposes for using media.
2. List three means of media that are non-technological. Describe the situations in which you would use them.
3. List two devices that you could use to develop an instructional program. How would you use them?
4. Describe the role of simulations.
5. Using a lesson plan, determine which media devices would be best to deliver the message.
6. You are assigned the task of selecting the media devices for your training center. Which devices would you select and why?

References

Hullfish, K. C. 1996. *Virtual reality monitoring: How real is virtual reality?* Unpublished master's thesis, University of Washington.

Rose, H. 1995. *Assessing learning in VR: Towards developing a paradigm virtual reality roving vehicles (VRRV) project.* Retrieved October 28, 2003, from *http://www.hitl.washington.edu/publications/r-95-1/.*

Bibliography

Brunacini, N. 2002. Command performance—Phoenix FD improves command response with new training center. *Fire Rescue Magazine, December,* 48–56.

National Highway Safety Transportation Administration. August, 2002. *National guidelines for educating EMS Instructors.*

Ross, F. 2002. Put the AI in training. *Fire chief, March,* 74–78.

Thiel, A., Stern, J., Kimball, J., and Hankin, N. 2003. *Trends and hazards in firefighter training.* (No. USFA-TR-100.) Federal Emergency Management Agency. Emmitsburg, MD.

CHAPTER 9

Curriculum Development

Terminal Objective

The participant will be able to design and develop a training course and lesson plan upon completion of this chapter.

Enabling Objectives

- List the five phases of the instructional design process.
- Describe the Analyze process.
- Describe the Design phase.
- Construct goals and objectives.
- Describe the Development phase.
- Describe the Implementation phase.
- Describe the Evaluation phase.
- Explain how a lesson plan is used.

JPR NFPA 1041—Instructor II

5-3.2 Create a lesson plan, given a topic, audience characteristics, and a standard lesson plan format, so that the job performance requirements for the topic are achieved, and the plan includes learning objectives, a lesson outline, course materials, instructional aids, and an evaluation plan.

5-3.3 Modify an existing lesson plan, given a topic, audience characteristics, and a lesson plan, so that the job performance requirements for the topic are achieved, and the plan includes learning objectives, a lesson outline, course materials, instructional aids, and an evaluation plan.

5-5.3 Develop a class evaluation instrument, given agency policy and evaluation goals, so that students have the ability to provide feedback to the instructor on instructional methods, communication techniques, learning environment, course content, and student materials.

JPR NFPA 1041—Instructor III

6-3.2 Conduct an agency needs analysis, given agency goals, so that instructional needs are identified.

6-3.3 Design programs or curriculums, given needs analysis and agency goals, so that the agency goals are supported, the knowledge and skills are job related, the design is performance based, adult learning principles are utilized, and the program meets time and budget constraints.

6-3.5 Write program and course goals, given job performance requirements and needs analysis information, so that the goals are clear, concise, measurable, and correlate to agency goals.

6-3.6 Write course objectives, given JPR's, so that objectives are clear, concise measurable, and reflect specific tasks.

6-3.7 Construct a course content outline, given course objectives, reference sources, functional groupings and the agency structure, so that the content supports the agency structure and reflects current acceptable practices.

6-5.3 Develop a course evaluation plan, given course objectives and agency policies, so that objectives are measured and agency policies are followed.

6-5.4 Create a program evaluation plan, given agency policies and procedures, so that instructors, course components, and facilities are evaluated and student input is obtained for course improvement.

Trainer Tales

The training fire set in a bus had been allowed to preburn for 10 minutes. While the crew was inside, conditions degenerated into flashover, and one firefighter was trapped after a disorganized exit by the others. He went into respiratory arrest as he was removed from the bus. He was revived but spent over two months in a critical care burn unit. Two others were less seriously burned but required weeks of treatment in burn units.

One of the major lessons of this incident was the recognition of the inappropriate nature of such improvised structures (in this case a modified school bus) for such a hazardous operation. The near-total lack of organization and safety procedures is a major lesson to learn from this unfortunate incident.

◆ INTRODUCTION

The accompanying Trainer Tales illustrates the concern for safety of students during training exercises. (See Figure 9.1.) At times, instructors must develop their own curriculum regardless of the many excellent courses on the market, or those that another instructor has developed and delivered successfully. The fire service is not a stagnant field. New innovations, new techniques, or different agency services emerge. If the instructor determines a need to develop a training course or program (which will be referred to as *curriculum*) to meet these needs, this chapter will go

FIGURE 9.1 ◆ Live fire training simulators can be a mobile unit. Using a simulated fire trailer is much safer than using an acquired structure and conducting a live fire burn. *(Photo courtesy of James Clarke, Estero Fire Rescue)*

through the process. In addition, this chapter will discuss writing course objectives and developing a lesson plan.

◆ CURRICULUM DEVELOPMENT

To efficiently and effectively develop a training course that will be successful, a number of sequential steps need to be followed. The Instructional System Design (ISD) process is the basis of a number of models that are used to develop a sound training course. This chapter will use the ISD process.

There are five phases to the ISD process:

- Analyze
- Design
- Develop
- Implement
- Evaluate

ANALYZE

The first step in the process is *analyze*. The key points in this step are:

- Analyze the system (department, job, etc.) to gain a complete understanding of it.
- Compile a task inventory of all duties associated with each job.

- Select tasks for training (needs analysis/needs assessment).
- Build performance measures for tasks to be trained.
- Choose the instructional setting for tasks to be trained: for example, classroom, on-the-job, self study, etc.
- Estimate the cost to train these tasks.

Let's examine these points in order.

Analyze the System

Curriculum development begins with having a complete understanding of the total system. This typically comes easily for the fire service instructor who is developing curriculum for department personnel. In these instances many instructors have been an integral part of the system and have a very good understanding of all the components. In other settings the fire service instructor needs to gain a thorough understanding of the system and not take anything for granted.

Compile a Task Inventory

It's important to first acquire the organization chart of all the departmental positions and how they fit into the department. Then review each of the job descriptions and identify what each position is responsible to perform. If neither of these items is available, the NFPA standards have Job Performance Requirements (JPRs) associated with various positions in the fire service. For example: Each of the chapters in this text has JPRs as they relate to the fire service instructor. When developing a program, an instructor needs to coordinate the objectives and the material to the JPR, whether using the NFPA standard or other standards to guide the development.

How to determine what a task is?

- A task has a beginning and an end.
- Tasks are usually measured in minutes or hours.
- Tasks are observable. This can be accomplished by observing the performance established in the JPR.
- Each task is independent of other actions. Tasks are not dependent on components of a procedure.
- A *task statement* describes a highly specific action. It always has a verb and an object. It may have qualifiers, such as "calculates flow rates without the use of a calculator." A task statement should not be confused with an *objective,* which has conditions and standards.

Job descriptions will list duties describing what needs to be performed. Duties are a combination of related or like tasks. For example, a fire service instructor might have the following two duties:

1. Create a lesson plan.
   a. Given a topic, audience characteristics, and a standard lesson plan format, create a lesson plan so that the job performance requirements for the topic are achieved.
   b. Include learning objectives, a lesson outline, course materials, instructional aids, and an evaluation plan.
2. Modify an existing lesson plan.
   a. Using a given topic, audience characteristics, and a lesson plan, modify an existing lesson plan so that the job performance requirements for the topic are achieved.
   b. Include learning objectives, a lesson outline, course materials, instructional aids, and an evaluation plan.

The duties listed above are developed from a fire service instructor's JPRs as listed by NFPA.

Select Tasks for Training

Needs Analysis. The next point to consider in the analyze phase is task or needs analysis. A task analysis is an intensive examination of how people perform work activities. The five steps to consider when conducting a task analysis are:

- Determine the components of competent performance.
- Determine activities.
- Determine sequence.
- Determine conditions.
- Determine performance standards.

Needs Assessment. A needs assessment should be conducted in the first phase before developing a training program. (See Figure 9.2.) A needs assessment could be comprised of any combination of the following methods:

- Observation
- Questionnaires
- Key consultation
- Print media

FIGURE 9.2 ◆ An instructor is conducting a needs assessment by interviewing a few of the individuals from their agency to determine the needs of the agency.

- Interviews
- Group discussion
- Tests
- Records, reports
- Work samples

The following should be considered when selecting tasks for training:

- Is the training mandated by any standard or regulation?
- Could a job performance aid or self-study packet be used in place of formal training?
- Can individuals be hired who have already been trained?
- Is training needed to ensure that employee behavior does not compromise the department's legal position; for example, Equal Employment Opportunity, labor relations laws, or state laws?
- What will happen if we don't train how to do this task?
- What will be the benefits if we do train how to do this task?
- If we don't train, how will the employee learn it?
- How will this training help to achieve our goals?

Build Performance Measures

Performance measures are the standards for how well a task must be executed. Four basic analysis techniques are used:

- Observation Task Analysis: Observing the task performed under actual working conditions and recording each step required and the standards of performance
- Simulated Task Analysis: Observing the task performed under simulated working conditions. In this situation the working conditions should emulate the job environment as closely as possible. Record each step and standards of performance with input from the skilled performers.
- Content Analysis: Analyzing the operating or technical manual to determine the steps and standards of performance
- Interview Analysis: A subject matter expert (SME) is consulted to determine the required steps and standards of performance. This is also common practice after the completion of the previous techniques.

Choose Instructional Setting

In Chapter 4 the importance of establishing an ideal learning environment was discussed. It's at this stage, when designing curriculum, to take the learning environment into consideration.

Estimate the Cost

It is essential to know the cost to develop the program or curriculum in order to establish a budget. In Chapter 10 budgets are covered in more detail.

DESIGN

The second phase of the ISD process is *design*. The key points of the design step include:

- Style guide
- Entry behaviors

- Learning goals and objectives
- Learning steps (performance steps)
- Performance test
- Program structure and sequence outline

Style Guide

A style guide needs to be established for the curriculum if one doesn't already exist. A style guide is used to maintain consistency and provide a guide for writing styles. This is especially important if there is more than one author writing the course and they are developing multiple curriculums. A useful guide to read is *The Elements of Style* by William Strunk, Jr. and E. B. White.

Entry Behaviors

Entry behaviors are what a learner must know before entering into the training program. An example would be requiring Chemistry of Fire before taking the Hazardous Material course. If you are instructing at the college level, certain standards must be met to enroll. Likewise, a training program requires a base level of knowledge, skills, and attitudes (KSA).

Learning Goals and Objectives

For instruction to have a strong foundation—and to meet the course intent—goals and objectives should be established for the educational materials. Aside from giving the instructor specific points to cover for each subject, they define for the students what they are to learn, and establish the parameters of areas to be tested and evaluated.

An entry-level instructor may not be asked to write objectives. However, instructors will work with educational curricula that contain objectives, and they must have an understanding of the basic components so that they will be able to determine if the objectives are meeting the teaching goals. In Chapter 7, writing test questions and how objectives assist in developing tests was discussed. Objectives can help determine how much information should be covered on a given topic plus the depth and breadth of the material you are presenting. This will separate what is "need to know" from what is "nice to know." Sometimes students need to master the material; other times they only need to be familiar with it. The objectives should help establish this.

Terminology. *Goals* are overarching, global statements of an expected learning outcome. An example of a goal statement would be: "The student will be able to use a hose line effectively." It is very broad and sets the mission of the topic, usually without any discussion of how to accomplish it.

Objectives are statements of expected learning in terms of behaviors. An objective should clearly state the expected behavior, the conditions it will be performed under, and what determines if the objective was successfully completed. A well-written objective should, in fact, lead to the completion of the goal.

TERMINAL AND ENABLING LEARNING OBJECTIVES

The Terminal Learning Objectives (TLOs) are a statement of the instructor's expectations of student performance at the end a specific lesson or unit. The TLO is written from the perspective of what the student will do—not what the instructor will do. The TLO consists of three parts: condition, task, and standard. TLOs are precise, observable, and measurable and stated in active terms. They typically represent a fairly large block of instruction, but will rarely range beyond a single lesson. The purpose of the TLO provides direction for a lesson. It forces the instructor to think through three questions:

What will the student be able to do as a result of completing the lesson?
Under what conditions (setting, supplies, equipment, etc.) will the student be required to perform the task?
How well must the student perform the task to pass?

Enabling Objectives (EOs) are concise statements of the instructor's expectations of student performance and might be considered steps in accomplishing the TLO. The EO is written from the perspective of the students and what they must do to accomplish the TLO. EOs typically provide only tasks and are observable and measurable, but often do not include the standard or condition. Each one involves a single step within the TLO. The purpose of Enabling Objectives is to specify a detailed sequence of student activities. It forces the instructor to think through the steps involved in completing the task in the TLO. The EO usually forms the outline for the instruction phase of the lesson plan. The instructor should think through a cycle of questions:

What is the first thing the student must be able to do (know)?
When that is complete, what must he/she do (know) next?

The chapters in this text have Terminal Learning Objectives and Enabling Objectives that can be used as examples for each.

BASIC PRINCIPLES

Goals and objectives need to be stated concretely with clearly identifiable and measurable outcomes. They should not be "fuzzy" or nebulous statements. For example, many objectives state that the learner will grasp a topic—say: A student will understand "hose lays." How do you measure this objective? Instead, the objective should state: The student will demonstrate how to perform a reverse hose lay. This is a measurable objective; you can easily determine if the student can indeed perform a reverse hose lay.

The curriculum designer needs to clearly communicate to the instructors and the students the expected behavior so that the goal can be accomplished. Each objective should relate to at least one goal, and each goal should be represented by at least one objective. Use the objectives to determine if the appropriate levels of content, depth, and breadth are being taught. How to accomplish this? Start with a preclass evaluation: Compare the lesson plan to what's written in the course goals and objectives. At the end of your class, review what was taught to determine if there were omissions. If the appropriate amount of information wasn't covered, there are two options. First, make sure the material is covered in the next class. If the class does not meet again, the second option is to provide an alternative learning mechanism—perhaps distribute a handout to the students; and of course, revise and enhance the lesson plan for the future.

COMMON CHARACTERISTICS OF GOALS

Think of goals as a vision or mission statement. Goals are global statements of intended learning. Goals may be philosophical in nature and don't communicate specific information on means of accomplishment or how to measure behavior or performance. An example: "The goal of this program is to provide the tools necessary to become an entry-level fire service instructor."

Goals are sometimes called primary objectives, first level objectives, or expected learning outcomes.

ABCD ELEMENTS

Many methods, models, and templates are available on writing objectives. One easy-to-remember method utilizes the letters ABCD to indicate the important information to include in an objective. (Note: A goal may or may not contain all the ABCD elements commonly seen in an objective.)

A = Audience, B = Behavior, C = Condition, D = Degree

An objective doesn't have to be written in this order (ABCD), but it should contain all these elements.

Two examples to follow in writing an objective: The (Audience) will _ (Behavior) in (Condition) circumstance to _ (Degree) level.

Or

Given _ (Condition) the (Audience) will _ (Behavior) to _ (Degree).

Audience

Describe your audience: Who will you be teaching?
Examples of audience statements:

The firefighter cadet
The company officer course participant
The firefighter attending your seminar

Behavior

Behavior describes what is expected of the student following instruction. The behavior must be observable and measurable. If it's a skill, it should be a real-world skill and relate to current fire fighting practices. The "behavior" can include demonstration of knowledge or skills in any of the domains of learning: cognitive, psychomotor, or affective.

Examples of behavior statements:

- Write a report.
- Assemble the equipment necessary to extricate a person from a vehicle.
- Defend the need to use forcible entry into a building.

Importance of Terminology

Wording like "should be able to" or "will be able to" carry different legal expectations and may be an issue to the organization. If you are writing objectives and are concerned about this, seek clarification from the supervisor or a senior instructor.

Condition

This describes any circumstance that can impact students' behavior. Equipment or tools that may (or may not) be utilized in completion of the behavior should be included. Environmental conditions or situations (temperature requirements, seasonal conditions, weather impact, swift water, time of day, etc.) may be included as conditions. Time limits might be imposed as a condition for performance. (See Figure 9.3.)

FIGURE 9.3 ◆ A firefighter is shown here with the instructor learning the procedure of using a SCBA. *(Photo courtesy of James Clarke, Estero Fire Rescue)*

Examples of condition statements:

Given a SCBA bottle, regulator, and mask…
Given the complete SOG manual…
Following donning SCBA and within 30 seconds…

Degree

The degree indicates the standard for acceptable performance (time, accuracy, proportion, quality, etc.). If the degree statement is not included in the objective, it's inferred that the acceptable standard for performance is 100 percent.

Examples of degree statements

Without error
Nine out of ten times
Without committing any critical errors

Review of ABCD Objectives

Well-written objectives will tell you the following:

Who is your target audience?
What observable performance is the student to exhibit?
What conditions are provided for the learner at the time of evaluation?
What constitutes a minimum acceptable response?

COMMON CHARACTERISTICS OF OBJECTIVES

Every objective should state how an expected behavior can be observed and describe how this behavior will be measured. (See Figure 9.4.) An example would be articulating that a particular psychomotor skill must be performed to a specific level of competency.

Another characteristic of objectives is that they are unambiguous. They should be written in plain language. Avoid jargon and define all terms the first time they are used. It should be clear to both student and instructor what behavior is expected to successfully complete the objective.

Objectives are results oriented. They are different from goals in that objectives describe specific expectations of performance, knowledge acquisition, feelings, or attitudes.

Objectives should be measurable by both quantitative and qualitative criteria. The easiest way to think of quantitative is to think quantity. Some examples of quantitative criteria include:

The lowest acceptable passing score
The number of attempts allowed during a skill test
A time limit imposed on a skill test

When thinking of qualitative criteria think quality. Qualitative criteria describe observations that express underlying dimensions or patterns of relationships. Some examples of qualitative criteria include:

Valuing a concept or idea
Defending the need to perform a skill
Adopting a new behavior

FIGURE 9.4 ◆ Objectives need to be established even when conducting practical evolutions.

One hundred percent accuracy on quantitative or qualitative measures is not required for every objective. An acceptable level of performance may be already established and allows the student to "miss" some elements but still pass the evaluation process. Here is an example:

> An acceptable minimum score for Firefighter I in a particular state is 70 percent, so an instructor requires all students to achieve a score of at least 75 percent on quantifiable objectives.

They may not have a required overall score for an objective. There may be items or steps identified as "critical criteria" that would result in failure if not performed. An example would be failure to use recommended PPE before performing a skill. The order in which the steps of the procedure are performed is as important as the steps themselves. An example of this would be not checking the amount of air in the SCBA before donning.

Objectives should be written in terms of performance and communicate successful learning in behavioral terms. If an objective doesn't describe or define the expected behavior, an instructor can't evaluate if learning has taken place. Some examples of expected behavior include:

- From an assortment of fire fighting equipment and supplies, select those items required to perform forcible entry.
- Demonstrate how to perform a database search on the Internet with a topic provided by the instructor.
- State three reasons why PPE should be used when performing salvage and overhaul procedures.

Examples of Objectives

- Given a standard sentence, the English 101 student should be able to correctly identify the noun and verb.
- The company officer participating in the incident command workshop should be able to identify the basic sectors that need to be established from a given scenario.

Trainer's Toolbox

Objective Specification Tool

Specifying program objectives as clearly as possible is a key element in instruction. Vague or poorly stated objectives can result in inappropriate instruction or invalid assessment. Consider the following objective: "Trainees will understand the importance of safety procedures on the job." This kind of vague objective provides little direction for how training should be designed or how the performance of this objective can be assessed. The objective specification tool assists you in the development and implementation of objectives in your coursework.

Performance objectives can be written for three domains: the cognitive, the psychomotor, and the affective. The cognitive domain refers to the intellectual processes involved in a job or content area. These can range from simple recall of information to complex problem solving. The psychomotor domain refers to skills that require coordination of the body in physical activity such as opening a container or moving a box. The affective domain refers to attitudes, beliefs, values,

Action Verbs for Writing Objectives

The following verbs are based on Bloom's taxonomy:

Know
Realize
Enjoy
Believe
Understand
Desire
Feel
Write

FIGURE 9.5 ◆ Action verbs that can be used to write objectives

and emotions such as enjoying or appreciating. The objectives specification tool helps instructors write more specific objectives by providing them with a list of observable verbs for both the cognitive and affective domains.

Instructions

Having precise performance objectives will assist in designing better interactions into the program and develop valid instruments to assess outcomes.

Writing performance objectives for the psychomotor domain is generally straightforward. For example: "The trainee will put on safety goggles and gloves before opening a container of any chemical substance." Writing performance objectives for the cognitive and affective domains can be much more difficult.

Whenever writing objectives, review the verbs in the accompanying box and in Figure 9.5 to select the ones most appropriate to the job or content. Though by no means a complete list, it is useful for writing better objectives.

Objectives Specification Tool

Level 1: ***Knowledge, or the ability to recall information.***

arrange	define	duplicate
label	list	match
memorize	name	order
recognize	recall	repeat
	reproduce	

Level 2: ***Comprehension, or interpreting information in one's own words.***

classify	describe	discuss
explain	express	identify
indicate	locate	recognize
report	restate	review
select	sort	tell
	translate	

Level 3: ***Application, or using knowledge in a novel situation.***		
apply	choose	demonstrate
dramatize	employ	illustrate
interpret	operate	prepare
practice	schedule	sketch
solve	use	

Level 4: ***Analysis, or breaking down knowledge into parts and showing interrelationships.***		
analyze	appraise	calculate
compare	contrast	criticize
diagram	differentiate	discriminate
distinguish	examine	experiment
inventory	question	test

Level 5: ***Synthesis, or bringing together parts of knowledge to form a whole and solve a problem.***		
arrange	assemble	collect
compose	construct	create
design	formulate	manage
organize	plan	prepare
propose	set up	synthesize
	write	

Level 6: ***Evaluation, or making judgments based on criteria.***		
appraise	argue	assess
attack	choose	compare
defend	estimate	evaluate
judge	predict	rate
score	select	support
	value	

Observable Verbs in the Affective Domain:		
agree	argue	assume
attempt	attend to	avoid
challenge	cooperate	defend
disagree	engage in	help
join	offer	participate
praise	resist	share
	volunteer	

BLOOM'S TAXONOMY OF EDUCATION

Domains of learning are based upon work done by educator Benjamin Bloom in the 1950s, which Bloom called the *Taxonomy of Learning*. Bloom's taxonomy of education that separates information into three discrete levels within each domain of learning provides a guideline for the six levels of the educational process. The domains are further divided into subsections that reflect the need for the students to have a deeper level of understanding (and sophistication) as they progress in each domain.

The first level of taxonomy is knowledge. This level includes objectives that deal with:

1. Specifics, such as specific facts and terminology
2. Ways and means of dealing with specifics, such as conventions, trends and sequences, classifications and categories, criteria and methodology
3. Universals and abstractions, such as principles, generalizations, theories, and structures

Example: The student will name the parts of an SCBA.

The second level of taxonomy is comprehension. This level involves objectives that deal with:

1. Translation
2. Interpretation
3. Extrapolation of information

Example: The participant will define a true emergency.

The third level is application. This level includes objectives that use abstractions in particular situations.

Example: The student will demonstrate proficiency on the driving course at each station. (See Figure 9.6.)

The fourth level of taxonomy is analysis. This level includes objectives that break the whole into parts and distinguish:

1. Elements
2. Relationships
3. Organizational principles

Example: The participant will differentiate between hepatitis B and hepatitis C.

The fifth level of taxonomy is synthesis. This level includes objectives that put parts together in a new form such as:

1. A unique communication
2. A plan for operation
3. A set of abstract relations

Example: The participant will be able to create an SOP/SOG.

The sixth level of taxonomy is evaluation—the highest level in terms of complexity. Objectives at this level would address making judgments using:

1. Internal evidence or logical consistency
2. External evidence or consistency with facts developed elsewhere

Example: The participant will be able to critique a patient care report.

FIGURE 9.6 ◆ A student is driving an engine through an EVOC course.

COGNITIVE DOMAIN

The cognitive domain emphasizes remembering or reproducing something already known about a subject. The three levels in this domain are:

- Level 1: knowledge (or recall), comprehension, and application
- Level 2: analysis
- Level 3: synthesis, and evaluation

PSYCHOMOTOR DOMAIN

The psychomotor domain is concerned with how a learner directs his or her body and emphasizes motor skill, manipulation of material and objects, or some act that requires neuromuscular coordination. The lower levels in this domain deal with skill performance with assistance or following a demonstration and progresses to "muscle memory," when the performance of the skill is done almost without conscious thought. The three levels in the psychomotor level are:

- Level 1: imitation and manipulation
- Level 2: precision
- Level 3: articulation and naturalization

AFFECTIVE DOMAIN

The affective domain is composed of two different types of behaviors: reflexive (attitudes) and voluntary actions/reactions (values). This domain is often difficult to write objectives and to evaluate if learning has taken place. Perhaps the best "teaching" that can be provided to the students in the affective domain is to model the behaviors you want them to adopt. The three levels in the affective domain are:

- Level 1: receiving and responding
- Level 2: valuing
- Level 3: organizing and characterizing

DOMAINS OF LEARNING: PLANNING LESSONS AND EVALUATING INSTRUCTIONAL TECHNIQUES

Before teaching, review the lesson plan and objectives to determine the depth and breadth of the material to cover for that session. After teaching, evaluate if the level taught was adequate for learning to take place. Here are some questions and examples to guide in this process.

- *Did it target the level specified in the objectives?*

EXAMPLE 1

The objectives state that the students should apply the information presented about communications by describing how they would react in a given scenario.

Class time was used to define terms, but no time was spent role-playing communications. Conclusion: The material was not taught to the level the students will be tested.

EXAMPLE 2

The objective states that the students should match a set of given terms to their correct definitions.

Students were given an extensive list of terminology, and the class was several hours behind schedule. Conclusion: The instructor went way beyond what was required by the objectives and threw off the schedule.

Activity

Write two objectives for each level of Bloom's taxonomy for a particular topic.

Learning Steps (Performance Steps)

Learning steps is the step-by-step process for conducting a class. The following is an example of this process.

1. Instructor receives class roster from training chief.
2. Instructor obtains necessary training documentation (e.g., lesson plan, course management plan) and supplies (e.g., learners' guides, slides, overhead transparencies) to conduct the class.
3. Instructor distributes the learners' guides prior to start of class.

4. Instructor checks the lab prior to class to ensure that all instructional items and equipment are present and in good working order.
5. Instructor arranges for delivery of any needed audiovisual equipment.
6. Instructor prepares for teaching role by rehearsing.
7. Instructor starts class on schedule.
8. Instructor presents the material listed in lesson plan and follows the general outline.
9. Instructor uses the following traits and techniques: flexibility, spontaneity, empathy and compassion, good questioning techniques, actively listens, gets feedback, positive reinforcement, and counseling.
10. Instructor directs the learners to find answers to their questions themselves rather than being an answering service.
11. Instructor provides coaching.
12. Instructor demonstrates new or difficult material in a comprehendible manner.
13. Instructor evaluates learners in the prescribed manner.
14. Instructor grades tests and distributes scores as required.
15. Instructor completes all learning activities and required functions during the allotted time period.
16. Instructor completes class roster and other forms at end of the training session and delivers them to the training department.
17. Instructor returns checked-out audiovisual equipment at the end of training session.
18. Instructor returns unused supplies and orders additional supplies if needed.
19. Instructor makes arrangements for the repair or replacement of damaged equipment.
20. Instructor ensures that the computer lab is in good condition for the next training session.
21. Instructor reviews the class just completed for new training ideas and then arranges to incorporate new training material into the lesson.

Performance Test

At this point in the curriculum design process, testing materials should be developed. (See Chapter 7 regarding testing and evaluation.)

Program Structure and Sequence Outline

The last step in the design phase is to determine the program structure and sequence to ensure the learning objectives are met. A proper sequence provides learners with a pattern of relationship so that each activity will have a definite—and obvious—purpose. The more meaningful the content the easier it is to learn and, consequently, the more effective the instruction.

Trainer's Toolbox

If a course is designed for a client, make sure the client puts his/her fingerprints on the document throughout the process. Nothing is more frustrating than putting a lot of time and effort into developing a program and then finding out it isn't what the client wanted. One way to monitor the entire process is to use a tracking system. Figure 9.7 is a sample of a curriculum tracker I've used that allowed me to track where I was in the process and the time line I needed to follow.

Curriculum Tracker

Program:

SME:

Project Manager:

Step	Date Due	Date Rec'd	Hours	Comments
Initial Review and Approval				
SME Content Draft Due to Curriculum Designer				
Curriculum Designer Format Complete				
SME, Dept. Review, Corp. Review Complete				
Revised by SME				
Final Format and Review by Curriculum Designer				
Professional Editing				
Graphics				
Final Review				
Pilot Program				
Final Revision				
Blue Line				
FINAL Edit				
Print Completion				
Distribution				

FIGURE 9.7 ◆ An example of form that can be used to keep track of the progress of the curriculum development

DEVELOP

The third step in the ISD process is *develop*. The development step includes:

- List activities that will help students learn the task.
- Select the instructional delivery method.

- Review existing material so that you don't reinvent the wheel.
- Develop the instructional courseware.
- Synthesize the courseware into a viable training program.
- Confirm that the instruction accomplishes all goals and objectives.

Let's look at these steps in order.

List Activities That Will Help Students Learn the Task

This step of the process is listing activities that will help the student best comprehend the material and learn the task at hand. Refer to Chapter 3 for review of the different ways individuals learn.

For example, if teaching a class on pump operations, the activity list may look like this:

Pump Operations

Classroom
Workbook on hydraulics
Hands-on
On the job

A number of different methodologies can be used to teach a subject. It's essential at this step to identify those ways. Some examples to consider are

- Classroom
- Lecture
- Video
- Workbook
- Computer-based training
- Job aid
- On-the-job training
- Coaching
- Mentoring

Select the Instructional Delivery Method

In Chapter 8 the different forms of instructional media were discussed. The type of media used to present the material—as well as whether to use existing media or to develop media to correspond with the topic—needs to be decided at this stage. Just remember that it's essential to keep the cost of the media within the budget established for the program. Developing media becomes very costly and adds additional time to the development of curriculum in many instances. This needs to be taken into consideration when deciding on a delivery method.

Review Existing Material so That You Don't Reinvent the Wheel

The terms "canned," "merit badge," and "alphabet soup" are sometimes used when referring to courses that have been developed for the masses. There is no disrespect for these courses; in fact, they're a resource for the instructor. Instead of having to go through the process of developing a course, an instructor can use one that already exists. Using a proven preexisting course is much better than developing one. Just keep in mind that these courses have been developed for large groups and may need to be fine-tuned to meet the needs of the class.

An example applicable to the fire service is the emergency vehicle operator's course (EVOC). EVOC courses are required by many states, insurance companies,

FIGURE 9.8 ◆ Pictured is a driver completing an EVOC course driving an ambulance.

and agencies; and a number of EVOC programs are available. If assigned to teach an EVOC course, it's probably not worth the time to develop a course—plus developing an EVOC course carries a liability issue with it. (See Figure 9.8.)

The "canned" EVOC course has all the relative material needed to teach and certify the students. The exception is that the course may not be designed with the pertinent information that applies to "your" personnel and their needs. For example, the laws that pertain to the state the students work in may not be part of the course package. Nor will the organization's SOGs be part of the course package (Figure 9.9). In these situations, most training programs encourage an instructor to bring this information into the course to make it applicable to the students.

Develop the Instructional Courseware

Now that all the planning has been accomplished, it's time to construct the instructional material. Developing different forms of the course requires a certain amount of skill and art. A number of different models can be used to accomplish this. The two that will be examined in this chapter are Keller's ARCS model of motivation and Gagne's nine instructional events (see also Chapter 3).

These nine events are a good guideline for developing the course or curriculum.

1. Gain attention.
   Examples:
   - Storytelling
   - Demonstrations
   - Present a problem to solve.
   - Show how to do something the right way after it was demonstrated the wrong way.
   - Explain why it is important.

FIGURE 9.9 ◆ Students should be familiar with SOGs.

2. Inform the learner of the objective.
   Example: Tell the person what you are going to tell them, tell them, and tell them what you told them.
3. Stimulate recall of prior knowledge.
   Example: Relate the current lesson to prior knowledge.
4. Present the material.
   Example: Chunk the information to aid memory and not overload the student. Bloom's taxonomy is a good guideline to accomplish this step.
5. Provide guidance for learning.
   Example: This is instruction on how to learn, not on the course content.
6. Elicit performance.
   Example: Allow the student to practice what they have learned.
7. Provide feedback.
   Example: Refer to the evaluation and feedback sections of this text.
8. Assess performance.
   Example: Test to determine if the objectives and the material have been learned.
9. Enhance retention and transfer.
   Example: Provide additional practice and on-the-job mentoring.

According to John Keller, there are four steps in the instructional design process: Attention, Relevance, Confidence, and Satisfaction (ARCS).

Attention. Gain students' attention by: perceptual arousal that uses the act of surprise or uncertainty, and inquiry arousal that stimulates curiosity by posing challenging questions or problems.

Examples:

- Use illustrations: visual stimuli, story, or biography
- Active participation or hands-on: role-playing, games, lab work, or other simulations
- Incongruity and conflict: play devil's advocate
- Inquiry: pose questions for them to solve
- Humor: keep in the proper context
- Variability: alternate presentation techniques

Relevance. Increase motivation by emphasizing relevance, using concrete language, and examples with which the learners are familiar. There are six major strategies for accomplishing this:

- Experience: Tell the learners how the new information will use their existing skills and improve their knowledge.
- Present worth: What will the information do for me today?
- Future usefulness: What will the information do for me tomorrow?
- Needs matching: Utilize Maslow's Hierarchy of Needs. (See Figure 9.10.)
- Modeling: Be the role model. Use others who act the part you are speaking.
- Choice: Allow the students to learn using the method they like best.

Confidence. Let the learners succeed! Do, however, present a degree of challenge that provides meaningful success.

- Provide objectives and prerequisites.
- Grow the learners; do not put them into information overload.
- Provide feedback.
- Help the learners feel they are in control.

Satisfaction. Give students opportunities to use newly acquired knowledge or skill in a real or simulated setting. Provide feedback and reinforcements to sustain the

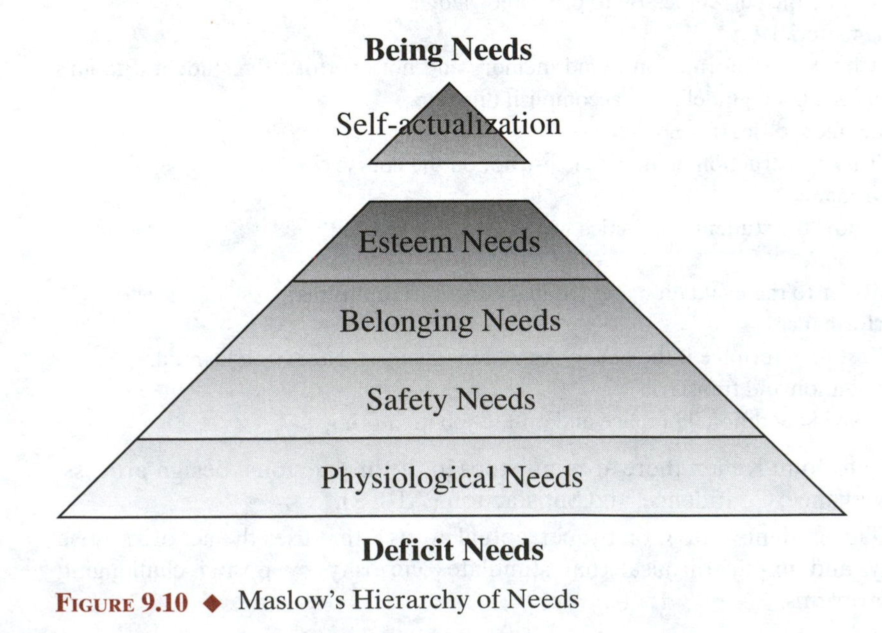

FIGURE 9.10 ◆ Maslow's Hierarchy of Needs

desired behavior—feeling good about learning results is a motivation to learn. Satisfaction is based upon motivation that can be intrinsic or extrinsic. Some basic rules are

- Don't annoy the learner by overrewarding simple behavior.
- If negative consequences are too entertaining, the learners may deliberately choose the wrong answer. For example: If the class laughs at a student's incorrect answer or if a student's answers evoke an amused reaction, other students may answer incorrectly and reinforce this unwanted behavior.
- Using too many extrinsic rewards may eclipse the instruction.

Synthesize the Courseware into a Viable Training Program

The training material and media should be integrated into the program. The flow of the material should seem natural. Each course foundation should be a continuance or the next sequence of the previous foundation. Include variety in the instructional strategies, and include breaks throughout the program. Pay special attention to dividing the course in time blocks. In addition, the number of days and hours per day will need to be determined.

Confirm That the Instruction Accomplishes All Goals and Objectives

Once the program has been developed, it's a good idea to do an alpha test—essentially a dry run to make sure the course flows and isn't missing any critical components. After conducting the test, make any necessary revisions. Once this has been completed, a beta test is conducted with a sample group of the population for whom the course is designed. Keep in mind that the extent an instructor needs to go with this process will depend on the course being developed.

Trainer's Toolbox

Figures 9.11 and 9.12 show two forms to use to aid in the alpha and beta testing of a course.

Any revisions should be made to the program after the beta test. Once that has been done, the course is ready for implementation.

IMPLEMENT

The next step in the ISD process is *implement*. Implementing a training program takes two steps:

- Create a management plan for conducting the training.
- Conduct the training.

Create a Management Plan for Conducting the Training

The course management plan is implemented by ensuring that the course—materials, class environment, and instructors—is ready to go. The students are scheduled, and any preparation by the student and/or instructor is completed prior to the start

Curriculum Review

Program:
Contact Name: Phone: E-mail:
Reviewer:
Date: Due Date:
Hours: ________

Directions:
Please review the following program/manual, utilizing the criteria at the bottom of the checklist as your guideline. Spelling and grammar are essential to a well-written document. However, there is a process in our review for this correction. Please pay close attention to the language, the flow of the content, whether the content meets objectives, the accuracy of the content (please provide source of any content correction), cultural bias, reading level, technical terms and jargon, activities, and other comments. Use a red pen to mark directly on the page. Please send the entire document back to ___________ by the date noted above. It is appreciated if you tag the pages with comments.

Item	Overall Comments
Language	
Content Flow	
Objectives met	
Accuracy	
Cultural Bias	
Reading Level	
Technical Terms	
Activities	

Please note any other comments on the back of this sheet.

FIGURE 9.11 ◆ Example of form to use to have others review the curriculum you develop

of the course. A train-the-trainer class may be needed to teach the instructors the new program.

A training management plan should be developed and contain the following information:

1. A clear and complete description of the course
2. A description of the target population
3. Directions for administering the course
4. Directions for administering and scoring tests
5. Directions for guidance, assistance, and evaluation of the learners
6. A list of all tasks to be instructed
7. Course map or course sequence
8. Program of instruction—how the course is to be taught
9. A copy of all the training material; for example, training outlines, student guides, etc.
10. Instructor and staff training requirements (needed and accomplished)
11. Any other documents related to the administration of the course

Curriculum Review Guide

Program Title ______________________

Reviewer ______________________

Criteria	Excel.	Fair	Poor	Comments
Layout/Organization	_____	_____	_____	—
Program structure (organization into sections, subsections, etc.)?	_____	_____	_____	—
Figures and tables (clearly labeled and professional looking)?	_____	_____	_____	—
Clearly stated purpose?	_____	_____	_____	—
Program accomplished stated purpose?	_____	_____	_____	—
Good overall structure? Ideas ordered effectively?	_____	_____	_____	—
Transitions used effectively?	_____	_____	_____	—
Introduction and conclusion focus clearly on the main point?	_____	_____	_____	—
Points developed in logical sequence?	_____	_____	_____	—
Each paragraph unified, developed, and coherent?	_____	_____	_____	—
Paragraphs right length for reading (not too long or too short)?	_____	_____	_____	—
Topic sentences included appropriately?	_____	_____	_____	—
Development and Support	_____	_____	_____	—
Major ideas/topics received enough attention and explanation?	_____	_____	_____	—
Supporting material persuasive?	_____	_____	_____	—

(*Continued*)

FIGURE 9.12 ◆ Form that can be used to review and evaluate curriculum

Adequate references and resource material?	______	______	______	—
Unnecessary repetition and redundancy avoided?	______	______	______	—
Style	______	______	______	—
Topic and level of formality appropriate for audience?	______	______	______	—
Sentences and words varied?	______	______	______	—
Wordiness avoided?	______	______	______	—
Grammar and Mechanics	______	______	______	—
Grammar?	______	______	______	—
Spelling?	______	______	______	—
Punctuation?	______	______	______	—
If you could recommend three specific changes in the writing, what would they be? 1. 2. 3.				

Figure 9.12 ◆ *(Concluded)*

Conduct the Training

The curriculum designer's role has essentially ended, and the instructor's role has begun. (See Figure 9.13.)

EVALUATE

The next step in the ISD process is *evaluate*. Evaluating a program means:

- Review and evaluate each phase (analyze, design, develop, implement) to ensure it accomplishes its objective.
- Perform external evaluations: Observe that the tasks that students were trained to do can actually be performed on the job.
- Revise the training system to make it better.

Review and Evaluate Each Phase (Analyze, Design, Develop, Implement) to Ensure It Accomplishes Its Objective

Everyone in the training system is charged with this step: Focus on the instructional processes, and what was gained from the training program. The primary purpose is to determine whether the instructional development accomplished what was intended. Enough data must be collected so that through time the instruction can be improved

FIGURE 9.13 ◆ Once the curriculum has been developed, the course is delivered in the classroom. *(Photo courtesy of Don Abbott, Phoenix Fire Department)*

based upon learner performance. If a large percentage of learners have trouble with the same segment of instruction, it's reasonable to conclude something is wrong with the instruction. (Refer to Kirkpatrick's four levels of evaluation as described in Chapter 7.)

Perform External Evaluations: Observe That the Tasks That Students Were Trained to Do Can Actually Be Performed on the Job

After the internal evaluation has been completed, one major question about the entire training program remains unanswered: Can the learners do the job they were trained for? The entire training process is designed toward this end. If the participants who complete the course don't need what they were taught or need additional instruction, this information needs to be feed back to the curriculum designers.

The various instruments used to collect the data are questionnaires, surveys, interviews, observations, and tests. The methodology used to gather the data should be a specified step-by-step procedure, carefully designed and executed to ensure accurate and valid data.

Revise Training System to Make It Better

Once any training deficiencies have been noted, the ISD process is repeated to correct that deficiency. This does not mean that the entire training program is rebuilt, just the portions that aren't training the learners to standards.

Activity

Working with other instructors, develop a course utilizing the curriculum design process.

CREATING LESSON PLANS

The focus has been on the process of developing a new curriculum or training course, but in fact the majority of the time will be needed to create a lesson plan. The last section of this chapter will discuss the lesson plan.

LESSON PLANS

A lesson plan is an effective way to organize the presentation, providing a guide to follow. A lesson plan also assists in the evaluation process—remember, objectives determine the content of tests. The lesson plan and the objectives define the depth and breadth of the material to cover during class.

In many instances, a lesson plan may already exist for a course so it won't be necessary to prepare one; but it's important that to know and recognize the required components of a lesson plan. If using a prepared lesson plan, make sure it has all the appropriate elements. It is the instructor's responsibility to determine if it's complete for the class to be taught.

Purpose of a Lesson Plan

A lesson plan serves as the framework or guide to the instructor while the lesson is being presented. It should assist the instructor in the selection of content to be presented. The lesson plan may contain notes about areas of importance that need to be emphasized or discussed, but a lesson plan does *not* take the place of preparation.

Sources of Prepared Lesson Plans

The state fire office may have predeveloped lesson plans available. Many federal agencies have lesson plans that accompany the programs, and an instructor can get them in a number of places. Publishers are another source of lesson plans. Many textbooks have an instructor's guide or manual that accompanies the text. In most instances, you must ask the publisher for a copy of the guide that typically mirrors a lesson plan. Organizations with certification and continuing education courses often have lesson plans that accompany their courses. The International Society of Fire Service Instructors publishes a different lesson plan every month in a monthly publication. Ask other instructors for lesson plans that they have developed. Remember, though to review these plans to make sure they conform to the required components of the course. With a bit of research, the wheel may not need to be reinvented.

Trainer's Toolbox

Sample Lesson Plan Outline

Audience description
Pertinent needs assessment information and prerequisites
Lesson goal(s)

Cognitive objectives
Psychomotor objectives
Affective objectives
Recommended list of equipment and supplies
Recommended schedule
Suggested motivation activity
Content outline

Needs Assessment

Needs assessment is performed before a lesson plan is written in the same manner as when you develop a curriculum. The anticipated training is evaluated to determine *who, what, where,* and *when.*

Who will attend the course? Identify the audience. Determine the demographics of typical and atypical students. Course content may affect various groups differently. The age and the experience of the student are also factors. (See Figure 9.14.)

The students may be coming from various distances that may be a factor to the start and stop times. Take into consideration weather and traffic issues. If teaching an all-day class, it may be beneficial to break for lunch early to allow students to beat the lunch crowd at local restaurants.

The motivation between volunteer versus career (paid) firefighters may be different. Although each group is comprised of professional students, motivations (intrinsic

FIGURE 9.14 ◆ Today's firefighters are very diverse. There are different cultures, genders, and ages in the fire service. The fire service instructor needs to take all these factors into consideration. *(Photo courtesy of James Clarke, Estero Fire Rescue)*

and extrinsic) may be different between a volunteer and a career participant. The biggest question to ask is: "Are they required to be there, or do they want to be there?"

Learning preferences and styles can vary greatly. (These were discussed in detail in Chapter 3.) Diagnostic instruments are available to determine students' preferences. Implement teaching strategies that will make learning more meaningful and enjoyable for the students.

Educational background is another consideration when conducting the needs assessment. Do students need additional preparation prior to entering the course? Who is responsible for providing the remedial or developmental education?

The prerequisites of the course need to be decided. Are there any entrance tests? What are the education prerequisites: English and/or math course work? Are there any certification level or experience requirements to take the course? Do participants need to show competency or performance verification prior to enrolling?

Technology requirements are another component. If technology is a component of the course, consider the impact of access to technology and user competence required.

The fire experience of the individual may be another key factor. What is the student's experience level? Are they taking the course to change careers?

Other commitments may detract from a student's learning capabilities. These may include: family and social, work schedules and responsibilities—shift work, inflexible schedules, or on-call status—and the time of day the class is offered conflicting with other commitments.

These all need to be taken into consideration when doing the needs assessment to create the lesson plan.

Equipment and Supplies

The lesson plan should list all the equipment or supplies needed to present the material. This should include: AV projection equipment, instructional equipment and supplies (easel charts, chalk, etc.), and any fire equipment and supplies. (See Figure 9.15.)

Recommended Schedule

A schedule guides the pace of the course. The class size and instructor-to-student ratios will affect the schedule. The physical location of the class will also affect the schedule. In Chapter 4 the classroom environment was discussed: poorly designed rooms, distractions, and poor temperature controls will affect students' concentration. Plan for frequent breaks. Some simple rules for breaks are:

1. Always plan a break within an hour following mealtime.
2. Break for at least 5 minutes each hour.
3. An optimal method is to vary the instruction at least every 20 minutes.
   Example: A 20-minute video clip followed by a 15-minute in-class exercise (then a 5-minute break) followed by a 20-minute lecture, a 20-minute skills demonstration, and then another break.
4. Plan breaks at appropriate times so that you don't interrupt momentum.

Use the Lesson Plan to Determine Content

The instructor must decide if the students need to have just a basic awareness of the material, or if they must master it. If unsure and there is a final exam, reviewing it may help determine how much material to present. Review Bloom's taxonomy to determine how detailed the presentation needs to be. The verbs used in the objectives will

FIGURE 9.15 ◆ Any fire equipment used for training needs to be prepared before the classroom session begins. *(Photo courtesy of James Clarke, Estero Fire Rescue)*

provide clues. As noted in the beginning of this chapter, cognitive domain verbs are placed into six groups from the lowest level required to the highest level of understanding. The six groups, in order, are knowledge, comprehension, application, analysis, synthesis, and evaluation.

The basic level of understanding is level 1. It includes objectives that demonstrate knowledge and comprehension. Students acquire new information or develop a new skill. This level requires feedback by the instructor.

Intermediate level of understanding is level 2. It includes objectives that demonstrate application. Students connect the knowledge learned in the basic level with knowledge gained through experience. For example: The instructor demonstrates the skill and then the student performs the skill with careful observation and correction by the instructor.

Advanced level of understanding is level 3. It includes objectives that require analysis, synthesis, and evaluation. The student functions with little or no supervision with the instructor serving more as a facilitator and coach than a teacher. The instructor focuses students toward learning why events occur as opposed to how to perform a skill.

Use a Lesson Plan to Present Course Content

The lesson plan can explain the importance of the curriculum. Begin with a statement listing and explaining the primary instructional goal and objectives. Allow the students to give feedback about the objectives. This is especially important when the audience is made up of professionals who have specific and intrinsic needs.

Let's look at delivering the content. Select methods suitable to student learning styles and limitations. Let the students practice the skills and document their competence level. Allow time for feedback from both instructor and student—encourage students to interact and contribute. Additionally, allow time for remedial education and evaluate the performance of the students and the lesson plan.

Student Tools

Encourage the students to take notes. It may be useful to provide students with an outline of the lecturer's notes. Several computer programs allow instructors to print a succinct outline of text and/or graphics for a presentation.

Promote interactivity in the classroom—perhaps have students submit questions during and after class time and encourage appropriate discussions. Encourage students to take responsibility for their learning. Don't spoon-feed them.

Evaluation of the Lesson Plan and the Presentation

Aligning curriculum objectives with specific lesson plan objectives is called performance agreement. The cumulative lesson objectives should address the course's goals. Lesson plans should build on previous course goals and objectives. Educational and practical objectives should support each other.

FORMATIVE EVALUATION

When writing the lesson plan, perform a formative evaluation. Compare the overall goal of instruction, the lesson objectives, and the content. Determine if there is a performance agreement between these three elements and make any adjustments necessary. If using a prewritten lesson plan, the evaluation is the same: Make sure that the instructional goal, objectives, and content are complete, and make any necessary adjustments so that there is performance agreement. Finally, review testing instruments to see if they match objectives and content.

SUMMATIVE EVALUATION

Summative evaluation is performed at the completion of the lesson to determine the effectiveness of your teaching strategy and improve future performance of the material. Methods of performing summative evaluation include: survey tools, test-item validation, and comparison of course and program outcomes.

Evaluation Tools

Tests and quizzes with the course objectives as their foundation are invaluable tools for documenting and evaluating student performance. Equally important is sharing the results of these evaluations in a timely manner. Have the student participate in designing a plan for improvement so that they can take ownership and responsibility for their progress.

Activity

Construct a lesson plan for a selected topic. Give it to another instructor to have them evaluate it.

◆ SUMMARY

Although a number of existing courses are available for the instructor to use, this chapter has provided the tools necessary to develop curriculum. Whether an instructor is using an already established course or creating a course, it is important to understand objectives and lesson plans. One of the first things a student asks when they come into a class is how long the class will last. Using a well-designed lesson plan, an instructor should be able to meet the objectives and goals established for the course in the time allotted.

Review Questions

1. Describe the process of curriculum design by listing the five steps and the process at each step.
2. List the five steps for a task analysis.
3. List at least five ways to conduct a needs analysis.
4. Discuss the positive features of using a "canned" course.
5. Discuss what the instructor needs to pay particular attention to when using a canned course.
6. Define goals, objectives, and performance agreements.
7. Describe the common characteristics of an objective.
8. Describe the components of an objective using the ABCD method.
9. List the six levels according to Bloom, and give at least three examples of a verb for each level.
10. List the three levels of the psychomotor domain.
11. List the three levels of the affective domain.
12. What is the purpose of the lesson plan?
13. List the components of a lesson plan.

References

Wiggs, G. 1984. Designing learning programs. In Nadler, L. (Ed.). *The handbook of human resource development*. New York: John Wiley.

Bibliography

Benefit, A. (1995). Instructional design process: A case example. *Performance & Instruction, September,* 40–42.

Bloom, B. et al. 1956. *Taxonomy of educational objectives, book I: Cognitive domain*. New York: Longman.

Butruille, S. 1998. "Lesson design and development." *American Society Training and Development, info line 8906.*

Carolan, M. 1993. "Seven steps for back-to basics training, nineties-style." *Training & Development, August,* 15–17.

Chapman, B. 1995. "Accelerating the design process: A tool for instructional designers." *Journal of Interactive Instruction Development,* (8)2, 8–15.

Clark, D. 1995. *Training handbook*. Retrieved January 15, 2004 from *http://www.nwlink.com/~donclark/hrd/sat.html#intro.*

Evers, L. 1992. "Designing an informational/instructional strategy." *Technical & Skills Training, November/December,* 25–31.

Filipczak, R. 1996. "To ISD or not to ISD?" *Training, March,* 73–74.

Ford, D., Ed. 1997. *ASTD's in action series: Designing training programs*. Alexandria VA: ASTD.

Gramiak, L. 1995. "Maintenance: The sixth step." *Training & Development, March,* 13–14.

Hodell, C. 1997. "Basics of instructional systems development." *ASTD Info-line, 9706.*

Holton, E., and Bailey, C. 1995. "Top-to-bottom curriculum redesign." *Training & Development, 49*(3), 40–45.

Huang, Z. 1996. "Making training friendly to other cultures." *Training & Development, September,* 13–14.

Kirkpatrick, D. 1996. *Evaluating training programs the four levels*. San Francisco CA: Berrett-Koehler Publishers, Inc.

Moller, L. 1995. "Working With Subject Matter Expert." *Techtrends, 40* (6), 26–27.

National Highway Safety Transportation Administration. August 2002. *National Guidelines for Educating EMS Instructors*.

Nooman, Z. M., Schmidt, H. G., and Ezzat, E. S. (Eds.). (n.d.). *Innovation in Medical Education*. New York: Springer Publishing Company.

Novak, J. D. 1977. *A theory of education*. Ithaca, NY: Cornell University Press.

Shultz, F., and Sullivan, R. 1995. "A model for designing training." *Technical & Skills Training,* January, 22–26.

Thiel, A., Stern, J., Kimball, J., and Hankin, N. 2003. *Trends and hazards in firefighter training*. (No. USFA-TR-100.) Emmitsburg, MD: Federal Emergency Management Agency.

Tracey, W. 1992. *Designing training and development systems*. (3d ed.) New York, NY: American Management Association.

Organizing and Running a Training Program

Terminal Objective

The participant will be able to organize and run a training program.

Enabling Objectives

- Describe how to plan and conduct a training course.
- Construct a training proposal.
- Explain how to budget for a training program.
- Describe the importance of training reports and records.
- Describe how to plan and schedule courses.
- Identify the key points to look for when using other instructors.
- Identify the key features of finding a training facility to conduct classes (see Figure 10.1).

JPR NFPA 1041—Instructor II

5-2.2 Schedule instructional sessions, given department scheduling policy, instructional resources, staff, facilities, and timeline for delivery, so that the specified sessions are delivered according to department policy.

5-2.3 Formulate budget needs, given training goals, agency budget policy, and current resources so that the resources required to meet training goals are identified and documented.

5-2.4 Acquire training resources, given an identified need, so that the resources are obtained within established timelines, budget constraints, and according to agency policy.

5-2.5 Coordinate training record keeping, given training forms, department policy, and training activity, so that all agency and legal requirements are met.

FIGURE 10.1 ◆ The Tarrant County Training Center outside of Fort Worth, TX

5-2.6 Evaluate instructors, given an evaluation form, department policy, and job performance requirements, so that the evaluation identifies areas of strengths and weaknesses, recommends changes in instructional style and communication methods, and provides opportunity for instructor feedback to the evaluator.

JPR NFPA 1041—Instructor III

6-2.2 Administer a training record system, given agency policy and type of training activity to be documented, so that the information captured is concise, meets all agency and legal requirements, and can be readily accessed.

6-2.3 Develop recommendations for policies to support the training program, given agency policies and procedures and the training program goals, so that the training and agency goals are achieved.

6-2.4 Select instructional staff, given personnel qualifications, instructional requirements, and agency policies and procedures, so that staff selection meets agency policies and achievement of agency and instructional goals.

6-2.6 Write equipment purchasing specifications, given curriculum information, training goals, and agency guidelines, so that the equipment is appropriate and supports the curriculum.

6-2.7 Present evaluation findings, conclusions, and recommendations to agency administrator, given data summaries and target audience, so that recommendations are unbiased, supported, and reflect agency goals, policies, and procedures.

6-5.2 Develop a system for the acquisition, storage, and dissemination of evaluation results, given agency goals and policies, so that the goals are supported and those impacted by the information receive feedback consistent with agency policies, federal, state, and local laws.

6-5.3 Develop course evaluation plan, given course objectives and agency policies, so that objectives are measured and agency policies are followed.

6-5.4 Create a program evaluation plan, given agency policies and procedures, so that instructors, course components, and facilities are evaluated and student input is obtained for course improvement.

Trainer Tales

In July 2002, a recruit died while participating in physical training. His death occurred on the third day of recruit school—the second day his class had participated in physical training. For their first physical training session, the recruits ran 2.78 miles. The next day, their physical training included a run of over 4 miles, a session of calisthenics, wind sprints, and jumping jacks. The heat index that morning ranged from 80° F at 7A.M. when training began to 99° F at 8:25 A.M. at the time the recruit was en route to the hospital. He was pronounced dead an hour later, with the cause of death listed as hyperthermia.

The department's board of inquiry found the following circumstances had led to the recruit's death:

- The recruits had no opportunity or means to hydrate themselves during physical training.
- Only one instructor was responsible for overseeing the recruits on the morning of the incident—an unacceptable ratio of instructors to students.
- The instructor leading the physical training session had no portable radio or cell phone to use in case of emergencies.
- Activating an EMS response was delayed because the recruit's condition was not initially recognized as a true medical emergency.

◆ INTRODUCTION

The years 1987 to 2001 saw a 31 percent decrease in the incidence of structure fires throughout the United States. This decline results in firefighters on the whole having less fireground experience than their predecessors had a generation ago. A comprehensive training program is essential for the success of training and educating the new generation of firefighters.

Depending on the size of the training program, an instructor may be both the instructor and the program administrator. Though it's likely that the training program will have a program administrator or chief in charge and the instructor won't be responsible for the supervision of the program, there are areas of program administration that an instructor will need to handle.

◆ PLANNING AND CONDUCTING COURSES

For most training programs the instructor is responsible for planning and conducting the course. A popular saying states, "If you fail to plan, you plan to fail." This is true with your training program. The secret of running a successful training program can be described in three words: plan, conduct, and evaluate. (See Figure 10.2.)

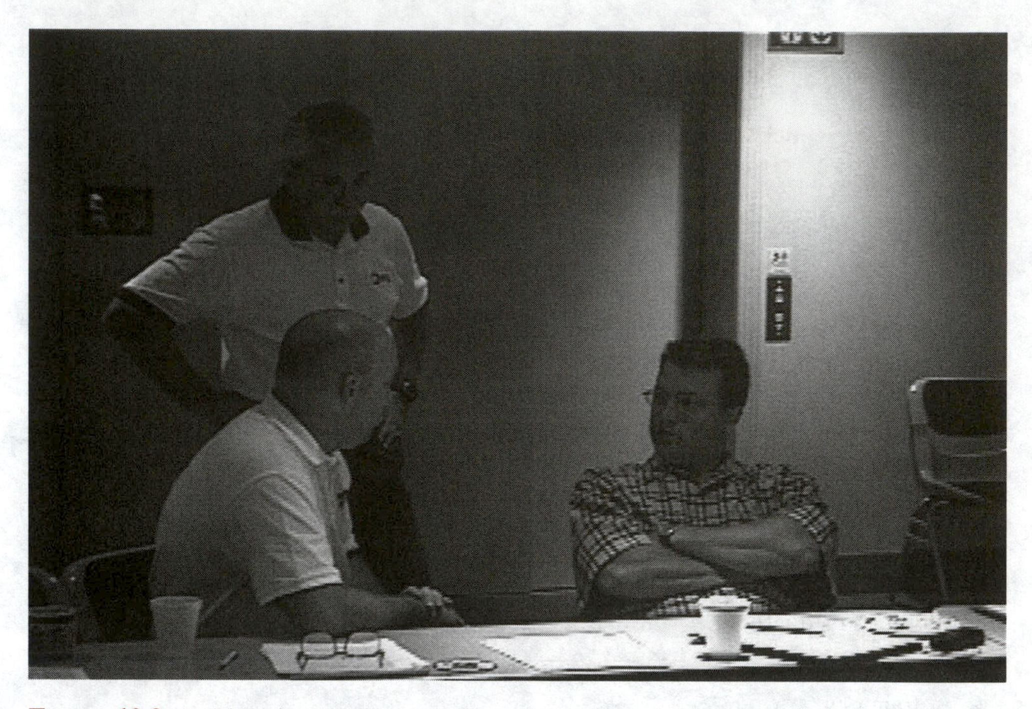

FIGURE 10.2 ◆ Planning sessions are important to chart the course of the training center.

This text has covered a variety of issues that need to be considered in the instructor role. Let's recap some of the elements of a good training course/program. It should:

- Provide for participants to draw on their own knowledge and experience.
- Encourage reflection.
- Encourage interactions based on openness, honesty, and respect.
- Seek to find answers to problems.
- Use real examples and situations for illustrations.
- Use various methods and approaches.
- Have established objectives.
- Have content that is linked: Each module builds on the previous one.
- Be relevant to the participants' work.

This section gives an overview of a systematic process that will enable the instructor to easily and successfully plan any type of training program. The first consideration is to determine what type of program to plan. As a guide, answer the following questions.

- Is it for a long or short term?
- Who is the intended audience?
- Why are you developing the program?
- How is the program to be conducted?
- Is this a new or existing program? If it is an existing program, what is the past history of its success or problems?

The next consideration is to set goals for the program. The following questions should be answered.

- What are you trying to accomplish?
- What will a participant learn or gain?

After answering these questions, use this information in planning, promoting, and evaluating the program. (The evaluation method needs to be determined; see Chapter 7 for evaluation techniques.)

Another consideration is to divide the tasks of the program into subtasks based on the course objectives. Answer the following questions for this step, and then determine realistic time lines to accomplish each task. Build some flexibility into the time line, but make every effort to keep the limits established.

- How many people are needed to conduct the program?
- What is the budget for the program?
- What type of facility is needed?
- Do you need to make arrangements for transportation, lodging, or food?

Depending on the size of the program being planned, a planning team may be needed that is made up of the key players. Anyone who is involved in the process should be included as part of the team. If delegating responsibility, it is important to also delegate the authority. The team needs to be clear on the expectations of the program. It is important to monitor the progress and provide any assistance as needed.

Another essential is determining what funds, supplies, and equipment are needed. First, establish a budget for the program if that hasn't yet been done. (Budgeting is discussed later in this chapter.) Then make a list of supplies and check to see if the necessary equipment is available or if it needs to be bought, borrowed, or rented. The number of people who are planning to attend is important. It may be necessary to have advanced registration and if that's the case, a registration deadline may need to be set. Additionally, a minimum or maximum number of participants needs to be decided in advanced.

The second consideration is promoting the program. A variety of promotional methods are available, depending on the program and the audience. Perhaps post the program announcement on the Website, send out a mailer to different organizations, or advertise in a local publication. A public information officer (PIO), if there is one, may be able to provide assistance to the instructor. (See Figure 10.3.)

It's important to get things done on time, so a time line needs to be established for each step of the program. Once it's established, stick with it. The training environment is just like teaching a class: Start on time and end on time in everything you do.

The participants who will be attending the program have an expectation based on what they've read and heard about the program. It's important to deliver what is promised. Go the extra mile and make sure you overdeliver. It is better to underpromise and overdeliver versus to overpromise and underdeliver.

As part of the course development process, it's your responsibility to evaluate the program. The following is a laundry list of some of the items to consider at the completion of the course:

- Get feedback from the participants—verbally and in written form.
- How effective was the promotion of the program? Ask the participants how they found out about it.
- How well did the planning team work? What can be improved upon next time?
- Make sure all instructors and personnel are paid if that has been agreed on prior. And a thank-you note is always appreciated.
- Make sure the bills are paid for the class.
- Evaluate the budget to make sure the program cost stayed within it.

FIGURE 10.3 ◆ Promotional material is an excellent means to get the information out to potential students. Conferences do a good job of promoting their sessions.

◆ WRITING A TRAINING PROPOSAL

Many consultants present their clients with a training proposal to detail how they will accomplish the training. It is a good tool to use to ensure that each component of the training program is being met.

Following is a guide to writing a training proposal that covers the major components of a training program. Depending on the program being created, it may not be necessary to go to this degree of formalization or to include every component.

Title Page

The first page is the title page that should have the organization's logo, the name and date of the course, and the agency's name and address. If copyrighting the document, it should also be noted on the title page.

Table of Contents

Page 2 of your proposal is the table of contents. The TOC gives readers (including yourself) a quick reference to find items in the document.

Executive Summary

The executive summary is similar to an abstract. It should be no more than one page in length and synopsize the document. It needs to be well written, in the active voice, and powerful. This should also describe the program being proposed.

Background of the Problem

In Chapter 9 conducting a needs assessment was discussed. This is where the curriculum developer presents who performed the needs assessment, and what procedures were involved in the process. This section puts the proposal into the context of the problem to be solved. The stakeholders or those involved in the process need to be included here.

Analysis of the Problem

This section should tackle the problem and offer an explanation of why the training program is needed, and explain how it will bridge the gap between the identified problem and the learning required. The following items should be covered in this section:

- Who it will serve
- Solutions
- Options
- Causes
- Current conditions

Use charts, graphs, and other visuals to make the proposal both easier to read and to help convey your message.

Target Population

Address who will be served by—or benefit from—the training. The instructor should also include management's involvement as part of the description.

Rationale and Goals of Proposed Training

This section should ask and answer the question: What is the purpose of the proposed training, and what are the anticipated benefits? Use charts, graphs, and other visuals to illustrate this whenever possible. This section must demonstrate a strong defense for the proposed solution.

Competencies

This is where the objectives for the program are listed. This tells the reader exactly what the program is designed to teach the students.

Evaluation Strategies

This section describes how the program is to be evaluated, using Kirkpatrick's four levels of formative and summative evaluations (see Chapter 7). Any documents that are used to evaluate the program should be included in an appendix.

Overview of the Intervention

The overview broadly describes the training solution to the identified problem. Include the types of learning activities proposed, where they will take place, and who will be involved. Describing the nature of the training that will occur provides decision-makers with a picture of the training experience. Show sample programs, home page, and so on.

Curriculum Outline

Detail the content, organization, and sequence of the proposed training program. Be as thorough and as complete as possible.

Training Resources Required

This includes any instructional materials, hardware and software, and personnel that are needed. (See Figure 10.4.) Any work that needs completing by participants before the training event should also be described here as well as a list of handouts and other related materials.

Capabilities of the Instructors

A description of each of the instructors should be in this section. It needs to include their qualifications, references to other successful projects, and a list of satisfied clients.

Schedule Development

This should include a sequence of planning events that describes step by step what needs to be done, when it will be done, how long it will take, and in what sequence. The schedule needs to outline the stages necessary to complete specific and separate phases of the needs analysis and design of the training project plus provides an overview of the specific tasks to be done. It should show which tasks will be done se-

FIGURE 10.4 ◆ Training resources includes already-packaged training programs or textbooks designed for fire safety instruction.

quentially and which will be done in tandem with another activity. The development schedule charts the most expeditious path to design the training program. Use graphs, models, charts, and so on for clarity.

Delivery Schedule

The delivery schedule outlines when the pilot testing will be done, when any revisions will be done, and when the product should be available for presentation. The schedule may cover a time period as short as a few days or a year or more depending on the nature and strategic importance of the project.

Costs

List the expenses involved with the program, explaining them in enough detail to justify them. It's a good idea to provide some options. This allows organizations to select the proposal that fits their needs. Also, break up the cost by session and participant. This will identify the per-student cost of training. It also helps in establishing the budget for this and future programs.

Projected Benefits

This final section describes the productivity improvements, quality improvements, workplace improvements, and return on investment of the program. This will help to push the program, whether selling it for monetary benefit or selling it to superiors to include in the training program.

Appendix

Any statistical data, instructor résumés, reference lists, supporting materials, or other pertinent documents should be part of the appendix. Figure 10.5 is ideal to ensure that you have all the bases covered for your class.

BUDGETING EXPENSES

Some people hear the word budget and run the other way. Budgets seem scary and complicated if you don't understand the components, but constructing a budget is actually pretty easy.

There are two sides to a budget: an expense column and an income column. The expense column will include any costs, and the income column will detail the revenue. Keep in mind that a budget should be created for each training course. The overall training program will also have an overall budget.

If the training program is currently operational, you can save some time and effort by using the previous budget as a guide to determine the necessary budgetary components. If starting from scratch, the following are some areas of expenses to include. (See Figure 10.6.)

Salaries and Honoraria

This line item includes the costs of instructors, other course assistants, administrative support staff, and any evaluators. If paying these individuals for prep time, don't forget to include this amount. Another area that gets overlooked on this line item is benefits. If the person is an employee and not a subcontractor, there will be additional costs associated with the salary. At the very minimum, the employee will have FICA

Class Checklist

Prior to Class

Course Location:	____________	Course Date:	____________
Directions:	____________	Course Time:	____________
	____________	Contact:	____________
Instructor:	____________	Contact Phone No.:	____________

AV Equipment Needed		***Other Materials***	
TV	____________	Participant Manuals #	____________
VCR	____________	Handouts	____________
Slide Projector	____________		
Computer Projector	____________		
Computer	____________		
Overhead Projector	____________		

Class

Set Up Room	____________	Identify Smoking Areas	____________
Check AV Equipment	____________	Identify Rest Rooms	____________
Tear-out Evaluations & Course Certificates	____________	Identify Drinks/Foods	____________
Be sure to leave facility better than you found it	____________		

After Class

	Date
Check all paperwork	____________
Record-Keeping	____________

Notes: ________________________________

____

____

Figure 10.5 ◆ Class checklist

and Medicare costs. There may also be health insurance, uniforms, or other benefits as a cost.

Fees

A variety of fees may be associated with the course. Some courses also have fees for course approvals or the certification process. Additionally, if constructing a budget for a comprehensive training program, take into account fees for the training program's attorney, accountant, and insurance (liability, property, etc.).

Budget Process		
Budget Items	**Projected Cost**	**Actual Cost**
Staff Salaries		
Staff Benefits		
External Consultants		
Instructional Materials		
Facilities		
Food		
Travel		
Equipment		
Promotional Materials		
General Overhead		
Other		
Totals		

FIGURE 10.6 ◆ Sample template for a budget for a training program

Facilities

If there isn't a facility, classroom space will need to be budgeted. If the instructor has a training facility, other areas need to be included in the budget. (See Figure 10.7.)

The first area is the classroom. Any additional furnishings such as tables and chairs need to be considered. If running your own training program, the tables and chairs that

FIGURE 10.7 ◆ The administrative building at the Tulatin Valley Fire Training Center houses the offices of the instructors and has a variety of classrooms.

are currently used won't last forever; and it may be more efficient to budget the annual replacement of a certain number of chairs and tables rather than replace them all at one time.

Office space is another area to include in the facilities budget. If sharing certain office features, this will factor into the budget. Some items to keep in mind for the budget—both for the initial setup and ongoing costs—are desks, chairs, computers, telephones, answering machines, file cabinets, photocopier, fax machine, miscellaneous office equipment, office supplies, and cleaning supplies. Other areas necessary to budget for are utilities such as electric, heating and cooling, and water.

Materials

The instructor will need to budget for a variety of materials. First are the promotional materials to recruit students for the class—unless the instructor is doing an in-house program that won't need any promotional materials. If promoting a course, budget for flyers, letters, postage, and the like to relay course information and mail registration forms.

Another area to budget for materials is training aids. This includes teaching aids (blackboard, large tablet, overhead projector, computer, projector, TV, VCR, paper, pens, markers, etc.), training equipment (simulators, hose, etc.), disposable supplies, and cleaning supplies.

Of course, the budget needs to include course materials such as the syllabus, handbook, curriculum, records, handouts, instructor resources, and textbooks.

Finally, budget for any refreshments that are being provided.

Travel

If there is a need to account for any per diem expenses or mileage, this needs to be included. Plus travel to multiple sites, if any, have to be taken into consideration.

Income

Income usually has fewer line items than the expense column. Typically there is income from the course fees. Depending on the program, there also may be other funding, including grants, state funds, or other revenue.

Once the budget is completed, it should balance out with the income. Of course, it's always better to have more income than expenses, as this allows the opportunity to create a reserve fund for unexpected operating costs that may arise. If the expenses exceed income, this is more critical. What to cut or how to increase the income will need to be determined.

RECORD-KEEPING

Record-keeping is an important aspect of a training program. (See Figure 10.8.) OSHA insists that training records be maintained for three years from the date of training. The following information should be documented according to OSHA:

- The dates of the training sessions
- An outline describing the material presented
- The names and qualifications of persons conducting the training
- The names and job titles of all persons attending the training sessions

FIGURE 10.8 ◆ Record-keeping is an essential part of being an instructor. Every training center needs to maintain training records for a minimum of three years.

The true measurement of training performance is the training record. The basic objective in keeping training records is to record training in the simplest manner possible. Training records should regularly be examined as part of the quality improvement process. A variety of individuals may inspect these training records. When ISO does an inspection of a fire department for grading purposes, they will review the training records. If any type of legal action is being waged against the department, the training records may be subpoenaed to determine whether existing training requirements are adequate to prevent further injuries or damage.

NFPA 1401 is the recommended practice for fire service training reports and records and covers the following information on training records and reports:

Chapter 1 is the Introduction. In this chapter the scope, purpose, general information, and definitions are covered.

Chapter 2 is Elements of Training Documents. This chapter relays the five specific elements of information that need to be included in a training document: who, what, when, where, and why.

1. Who
   a. Who was the instructor?
   b. Who participated?
   c. Who was in attendance?
   d. Who is affected by the documents?
2. What
   a. What was the subject covered?
   b. What equipment was utilized?
   c. What operation was evaluated or affected?
   d. What was the stated objective and was it met?
3. When
   a. When will the event take place? or
   b. When did the event take place?
4. Where
   a. Where will the event take place? or
   b. Where did the event take place?
5. Why
   a. Why is the event necessary? or
   b. Why did the event occur?

Chapter 3 discusses the Types of Training Documents. It delineates the various types of training schedules and explains in detail the training report and training record. It concludes with State Certification Records.

Chapter 4 is Computerization of Reports and Records. It covers the general issues concerning computerized reports and records.

Chapter 5 is Evaluating the Effectiveness of the Training Records System.

Chapter 6 sets the standard for the Legal Aspects of Record-Keeping.

Chapter 7, the conclusion, cites Referenced Publications. The standard also includes a number of sample reports in Appendix B.

COURSE PLANNING AND SCHEDULING

Effective training requires ongoing instruction and efficient scheduling. The constantly changing environment and new techniques, rules, and regulations require firefighters to continually train to ensure that they can perform their duties in a safe manner. (See Figure 10.9.) Training can be classified into the following categories:

- Mandated training (required by local, state, or federal)
- Specialized training
- Annual training
- Quarterly training
- Monthly training
- Departmental training
- Needs assessment

Another way to categorize training is:

- Recruit training
- In-service training
- Special training
- Officer training
- Advanced training
- Mandated training

After categorizing all the training, instructors need to put the training schedule together. Depending on the geographic location, they may need to take into account the various seasons when planning the calendar. For example, if the training program is in the south, an instructor may not want to conduct a driver training program in the heat of the summer, when the vehicle may wreak havoc on asphalt. Likewise, if in a northern climate, an instructor probably won't want to conduct a driver training class in the winter and have to deal with frozen precipitation and freezing temperatures.

FIGURE 10.9 ◆ Careful planning needs to take into account environmental factors when conducting courses like EVOC.

Holidays and special events also need to be considered. Scheduling can be very challenging, given the time constraints in the fire service. Once the schedule has been established, post it so that everyone is aware of when the training events are scheduled. When inviting other personnel to the training event, make sure the dates and times are published well in advance and that everyone knows where the course is being conducted. If outside individuals are attending, directions to the facility need to be included.

◆ RECRUITMENT

When instructors reach the point that they are running a training facility, additional instructors will be needed to teach the courses. Recruiting instructors can be challenging; but after reading this book, an instructor should be well aware of what it takes to be an instructor. When recruiting for other instructors, refer back to those sections that discuss the role of the instructor and the qualities of an instructor. These are good to use in future recruiting.

One of the instructor's responsibilities as a training coordinator or manager is to evaluate the staff. Figure 10.10 is a tool that can be used to assess instructors. The completed form should be reviewed with the instructor, then filed in his/her personnel folder.

◆ FACILITIES

The instructor may not have much choice about where to do the training; the facility may be provided by the group being trained or held in the organization's training room. Whenever instructors use any facility they are unfamiliar with, they should visit the location beforehand to ensure it will be arranged to meet their needs for chairs, tables, and so on. (See Figure 10.11.)

When considering facilities for training, take these factors into consideration (Figure 10.12):

- Use broad rooms versus long and narrow rooms that feel more constrictive.
- Rooms need to be large enough to accommodate your largest classes with smaller breakout rooms nearby.
- Rooms need to have soundproof dividers.
- Be sure there is adequate power to run all conceivable AV equipment.
- Check for adequate lighting.
- The best colors for a room are cool pastels—grays, browns, creams.
- The clock should be on the back wall for the instructor.

No matter what and where the facility, always leave the room cleaner than you found it.

Activity

Put together a training program for a small fire department of 30 personnel. As part of this program, create a training proposal that includes a budget for the program. In addition, work up a training schedule that covers a year.

Evaluation of Instructional Techniques

Topic: ______________________________

Date: ______________ Evaluator: ______________

Instructor: ______________ Time: ________

Instructions: Grade each element by either an E = excellent; G = good; or NI = needs improvement. Please comment on each section.

Classroom Evaluation

Preparation for class ____

Communication of classroom expectation ____

Command of subject matter ____

Professional and businesslike classroom behavior ____

Lecture and discussion conform to topic ____

Encouragement of student participation ____

Comments: ______________________________

Course-Related Factors

Utilization of supplemental teaching aids ____

Effective utilization of personal stories ____

Comments: ______________________________

Instructor Personal Evaluation

Voice tone and projection ____

Personal presentation ____

Sense of humor ____

Comments: ______________________________

Instructor's greatest strength(s) ______________________________

Suggestions for improvement ______________________________

Figure 10.10 ◆ Staff evaluation form to be used by the lead instructor. The form is an excellent tool for critiquing new instructors.

Activity

You have been assigned to be the lead contact to set up and run a seminar on fire ground safety. Describe all the elements of setting up this seminar and complete all the necessary documentation to conduct it.

FIGURE 10.11 ◆ The Tarrant County Training Center

FIGURE 10.12 ◆ Careful consideration needs to be taken when designing the classroom environment.

◆ SUMMARY

Although this chapter took a cursory look at running a training program, keep in mind that an entire text could be devoted to this subject. A fire service instructor may one day be called on to run a training organization. If so, this chapter has given the basic understanding of what it takes.

Review Questions

1. What three words can be used to describe running a training program?
2. List the components that make up a training program.
3. Describe the process for conducting a training program.
4. Describe the process for writing a training proposal.
5. List the components of a budget, and describe the line items that would fit into each component.
6. Define the parameters of record-keeping, including the number of years to maintain training records.
7. Describe how to put together a training schedule.
8. Describe the characteristics you would look for in hiring an instructor.
9. What areas should you take into consideration when securing a training facility?

Bibliography

National Highway Safety Transportation Administration. (August 2002). *National guidelines for educating EMS instructors*.

Thiel, A., Stern, J., Kimball, J., and Hankin, N. 2003. *Trends and hazards in firefighter training*. (No. USFA-TR-100.) Emmitsburg, MD: Federal Emergency Management Agency.

CHAPTER 11

Professional Development and Resources

Terminal Objective

The participant will be able to develop their plan for professional development as a fire service instructor.

Enabling Objectives

- Describe the role of mentors.
- Identify various continuing professional development opportunities.
- Discuss the value of using a library as a fire service instructor.
- Describe research as it pertains to the fire service instructor.
- List various professional organizations.
- Describe various ways to obtain professional development opportunities.
- Describe Fire and Emergency Services Higher Education (FEHSE) and how it affects the fire service instructor.
- Discuss the benefits of Training Resources and Data Exchange (TRADE) to the fire service instructor.

JPR NFPA 1041—Instructor II

5-2.4 Acquire training resources, given an identified need, so that the resources are obtained within established timelines, budget constraints, and according to agency policy.

JPR NFPA 1041—Instructor III

6-2.4 Select instructional staff, given personnel qualifications, instructional requirements, and agency policies and procedures so that staff selection meets agency policies and achievement of agency and instructional goals.

INTRODUCTION

Fire service instructors must continue to expand their skills and abilities. This chapter is devoted to providing a number of resources for the new fire service instructor to use in professionally developing. The list is not exhaustive, but does provide a wealth of resources in both the fire and the adult education setting. (See Figure 11.1.)

PROFESSIONAL DEVELOPMENT

Continuous learning is a way of life. Reaching the level of fire service instructor does not mean your learning is complete. The best instructors never stop learning. In fact, the instructor sometimes learns more than the students when teaching. To be an effective instructor, professional development needs to be a part of your lifestyle. Professional development can come in many different shapes and sizes.

MENTORS

A mentor is a valuable resource to any instructor, not just a novice one. A good mentor will help direct continuing personal and professional development plus serve as a resource to help problem solve instructional issues.

FIGURE 11.1 ◆ There are a variety of training programs in fire service instruction around the country, ranging from a fire department training center to a university. The resources for professional development are abundant. *(Photo courtesy of Don Abbott, Phoenix FD)*

Mentors are also an excellent resource for contents, teaching methods, and techniques. A mentor provides some of the following:

- Guidance
- A good role to model
- Constructive criticism to help you grow (both personally and professionally)
- Insight from experiences

A mentor might be another fire instructor, an instructor from another public safety field, a former professor, or even a highly trusted and respected friend or relative.

As instructors progress through their careers in the fire service there are times when they will need various types of mentoring. Depending on what class that is being instructed, it may be beneficial to bring an individual into the class who works in the area that is being taught. For example, if teaching a fire department administration course and the topic is budgeting, bring the town's budget officer into the class. This adds credibility to the instruction and offers the students a potential mentor for when they do a budget.

Don't have tunnel vision for the fire service only. The industry is getting more complicated every day, and there is a need to begin to align with others who are the experts in their field. Students will look to the instructor for answers. In some cases, the answer is to direct them to a mentor.

CONTINUING PROFESSIONAL DEVELOPMENT OPPORTUNITIES

Conferences and seminars across the country are excellent opportunities for professional development. Current science is reviewed or presented as it relates to the fire service at most conferences. This is also a time for instructors to expand their background knowledge. Most conferences last two or three days and offer a variety of topics and sessions, which show how other individuals are doing certain evolutions, and it's also a great chance to meet other fire service professionals and establish a network of colleagues.

Some of the conferences cater especially to instructors and provide sessions on teaching methodology, course delivery, and development. This is also an opportunity to watch others teach, which in turn helps to make an instructor a better teacher.

Most conferences have an exhibit hall where vendors showcase their wares, providing an opportunity to see the newest products. In many cases if you say that you are an instructor, the vendor will offer free materials to use in the classroom. These companies need to make money, of course, so don't expect a lot of freebies; but they do provide some training materials for the classroom.

THE LIBRARY

The public library is probably one of the most underutilized resources—though just about every community has one. Generally easily accessible, libraries typically are very helpful in performing information searches, both traditional and online. Libraries also have meeting rooms that can be used for training sessions. (See Figure 11.2.)

Trainer's Tip

The regional public library in our fire district allows us to use a room for training and doesn't even charge us any user fees. Check out the local library and see what you may have been missing.

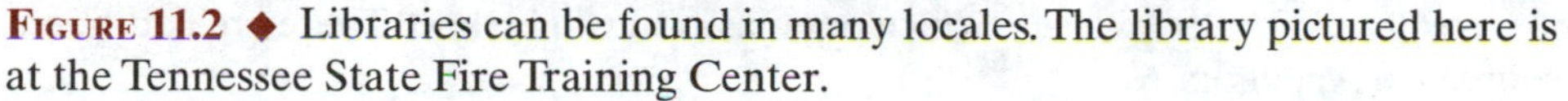

FIGURE 11.2 ◆ Libraries can be found in many locales. The library pictured here is at the Tennessee State Fire Training Center.

College Libraries

Academic libraries may have content-specific materials and access to more scientific material than a public library. If the local college teaches fire science courses, they are more likely to have fire-based texts. The staff at most college libraries, especially those that have fire-based resources, will be well versed in research strategies. Keep in mind that a college library may require that an instructor be a registered student, but the librarian might be able to work out a program if the instructor is with the local fire department.

A variety of general databases are available at both the public and private libraries, including CINAHL and NEXUS/LEXUS. ERIC is one of the most widely used database searches for educational literature. A number of libraries are linked together for interlibrary loans to make it easy for instructors to retrieve a text. Once instructors find the text that they want to use, they can find out in which library it's located and have the library send it to their local library. The National Fire Academy works in the same manner. Instructors can check for the publication on the USFA website and have it sent to their local library. When finished, take it back to the local library, and return it through the system.

RESEARCH

The fire service is unfortunately lagging in doing research. The industry needs to recognize that research is an important aspect of professional growth and development.

Research can be done to address a specific need for an organization. For example, the agency may want to know what the response times are for a certain station. Using research, a problem may be solved, a process changed, or training program developed. Research provides a basis for further study and future projects. Fire service professionals continually need to demonstrate the value to the community; this is one means to do this. Instructors need to demonstrate the effectiveness of the training.

The Learning Resource Center of the National Fire Academy in Emmitsburg, Maryland, is a comprehensive resource center where an instructor can access many items through the interlibrary lending system. There is also research available that has been conducted on fire-related subjects. Contact the center at 1-800-638-1821, or access the center's database through the NFA website.

PROFESSIONAL GROUPS AND ORGANIZATIONS

A number of professional groups for the fire service instructor provide mentoring and support both in the fire service and the adult education arenas. Also, many counties have a training officer group where training officers can exchange information and find out ways to do things. The American Society of Training and Development has local chapters in most regions; there, trainers from various disciplines get together monthly to discuss various ideas and learn from programs that are presented. This group may not understand fire-related issues, but they do understand instructional issues and can provide a significant resource.

Bottom line: Instructors can't belong to every group or organization so they should carefully select the organization or professional group that meets their needs. The websites may help to decide which organization will be best. There are additional resources in Appendix A.

Publications

Subscribing to magazines and journals—whether in traditional form or online—is probably one of the easiest ways to keep current on some of the technologies in the field. Just about every professional journal also has a website to get up-to-the-minute information on newsworthy events. Some of the most popular publications to check out are *Fire Engineering, Firehouse, Fire Chief,* and *Fire Rescue* (Figure 11.3). A number of these publications even have drills that can be incorporated into training programs, though a word of caution: Instructors need to make sure they fully understand how the drill works and its inherent dangers as noted in the accompanying Trainer Tales. Students have been injured—even fatally—as result of not understanding the drill prior to utilizing it during training.

Trainer Tales

An engine company was participating in regularly scheduled training evolutions at their academy when an officer suggested a maneuver he had seen in a trade magazine. The evolution involved escape from flashover conditions utilizing a ground ladder at an extremely low angle placed below the sill of the "escape" window. Permission was granted, and one of the firefighters attempted the maneuver that required that he exit through the window headfirst and climb down the ladder on all fours. The firefighter lost his balance and tumbled from the ladder, breaking a leg.

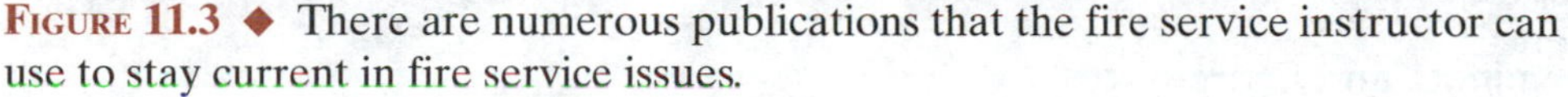

FIGURE 11.3 ◆ There are numerous publications that the fire service instructor can use to stay current in fire service issues.

While the initiative to try a new technique is highly commendable and should not be stifled, all training evolutions require careful consideration and verification that safeguards are in place. Further, the head-first ladder escape (or ladder rail/slide) is considered the escape method of last resort, and training on this technique should only occur in controlled situations. Additional information can be found on this through the USFA Technical Report, "Rapid Intervention Teams and How to Avoid Needing Them."

Professional Organizations

Join and get involved with professional organizations that will provide a network to share information with like-minded instructors. A few national organizations are the International Society of Fire Service Instructors, the International Fire Chiefs Association, and the American Society of Training and Development. Though this last organization is not a fire service organization, it provides a great network of training professionals and a wealth of useful information. Though space constraints limit listing all the professional organizations, check around. Numerous professional organizations would welcome a fire service instructor's involvement either on the national, state, or local level. One further suggestion: Consider joining one nonfire organization in the training field to give a different viewpoint.

FIGURE 11.4 ◆ Conferences have become more prevalent and are a great place to learn the latest information and network with you peers.

Seminars and Conferences

Seminars and conferences are another venue for professional development (Figure 11.4). Just about every national professional organization and magazine holds some type of annual conference and offers informational seminars. The largest instructor conference is Fire Department Instructor's Conference (FDIC), which has grown to include several locations. Other conferences are Fire Rescue, Fire Rescue International, and Firehouse Expo. Again, check around to see what seminars and conferences would be of professional and personal interest.

College Coursework

The most common source of professional development is taking a college course (Figure 11.5). Most fire service instructors have earned their associates degree, but that is not always the case. An associate degree should be the minimum goal of any fire service instructor. Most community colleges require instructors of technical programs to have an associate degree, though more often academic institutions want instructors to hold a bachelor's degree and many prefer a master's degree. A doctorate is not a common level of education in fire service instruction; however, depending on the institution and the level of instruction you are teaching, it may be a requirement. The higher the level of academic success you earn, the more opportunity you will have in the academic setting.

Make sure that any academic degree pursued is administered by an accredited institution, since a degree issued by a nonaccredited institution can be the same as not even having a degree. Many institutions require degrees to be earned from an accredited institution to teach at the college or university level. A number of accredited institutions offer degree programs through nontraditional means such as distance learning.

FIGURE 11.5 ◆ There are many universities and colleges that have fire science programs. Many of these have distance learning programs.

Fire and Emergency Services Higher Education (FESHE)

The National Fire Academy has taken an active role in higher education. The website is a source of valuable information about what is happening in fire service higher education today and in the future including the NFA's Degrees at a Distance Program, downloads from the annual FESHE conferences, the FESHE discussion forum, fire-related research sites, emergency management degree programs, scholarships, financial aid, and much more. To see what programs the NFA offers, go to: *www.usfa.fema.gov/fire-service/nfa/higher-ed/he.shtm.*

MODEL FIRE-RELATED ASSOCIATE DEGREE CURRICULUM

Since 2000 the Fire and Emergency Services Higher Education (FESHE) conference attendees have diligently moved to the forefront in developing and implementing the fire-related curriculum model.

One result of the conference was the development of the model fire science associate's degree curriculum. The FESHE attendees identified six core associate-level courses in the model curriculum:

- Building Construction for Fire Protection
- Fire Behavior and Combustion
- Fire Prevention
- Fire Protection Hydraulics and Water Supply
- Fire Protection Systems
- Principles of Emergency Services

In 2001, the National Fire Science Curriculum Committee (NFSCC) was formed to develop standard titles, descriptions, outcomes, and outlines for each of the six core courses. In 2002, the FESHE IV conference attendees approved the model courses and outlines. The major publishers of fire-related textbooks are committed to writing texts for some or all of these courses.

Fire science associate degree programs are encouraged to require these courses as the "theoretical core" on which their major is based. The course outline addresses the need for a uniformity of curriculum and content among the fire science courses in two-year degree programs. Many schools already offer these courses in their programs while others are in the process of adopting them. Once adopted, these model courses will address the need for problem-free student transfers between schools. Likewise, they promote crosswalks for those who apply their academic coursework toward satisfaction of the national qualification standards necessary for firefighter certifications and degrees.

The NFSCC developed similar outlines for other courses that are commonly offered in fire science programs. If a school offers any of these "noncore" courses, it is suggested these outlines be adopted as well. The noncore courses are

- Fire Administration I
- Occupational Health and Safety
- Legal Aspects of the Emergency Services
- Hazardous Materials Chemistry
- Strategy and Tactics
- Fire Investigation I
- Fire Investigation II

As fire science programs begin to offer these model courses and state training and certification agencies begin to recognize them, the professional development model can move from a concept to a practice.

BACCALAUREATE CURRICULUM

At FESHE IV in 2002, the NFA announced it would release its 13-course baccalaureate Degrees at a Distance Program (DDP) to accredited institutions that have signed agreements with their state's fire service training agency. DDP will remain as NFA's delivery system for the 13 courses; however, release to other schools enables the formation of model curriculum at this level. (See Figure 11.6.) The courses are

- Advanced Fire Administration
- Analytical Approaches to Public Fire Protection
- Applications of Fire Research
- Community and the Fire Threat
- Disaster and Fire Defense Planning
- Fire Dynamics
- Fire Prevention Organization and Management
- Fire Protection Structures and Systems Design
- Fire-Related Human Behavior
- Incendiary Fire Analysis and Investigation
- Managerial Issues in Hazardous Materials
- Personnel Management for the Fire Service
- Political and Legal Foundations of Fire Protection

FIGURE 11.6 ◆ The University of Cincinnati is one of the Open Learning programs that the National Fire Academy is involved with.

A NATIONAL SYSTEM FOR FIRE-RELATED HIGHER EDUCATION

With model lower-level (associate) curriculum outlines developed, and established upper-level (baccalaureate) courses available, the major components are in place to move toward a national system for fire-related higher education. This is important because, as with other professions, a theoretical core of academic courses should be a prerequisite for entering these fields. As more schools adopt these curricula, the fire and emergency services move towards becoming a full-fledged profession. (See Figure 11.7.)

A CALL FOR COLLABORATION

The relationship between the "big three" of the fire and emergency services professional development system—training, certification, and higher education—are varied across the country. In most states, levels of cooperation among the three range from low to nonexistent. Some exceptional state models of cooperation do exist including those of California and Oregon. The models' similarities demonstrate that:

- Partnerships can solve training, education, and turf battles by bringing together stakeholders in some formal or informal organization or consortium.
- Through cooperation a professional development delivery system that works for the state can be created and maintained.

From where must this leadership emanate? Leaders are needed at all levels. The state offices responsible for fire service training and certification, the fire-related

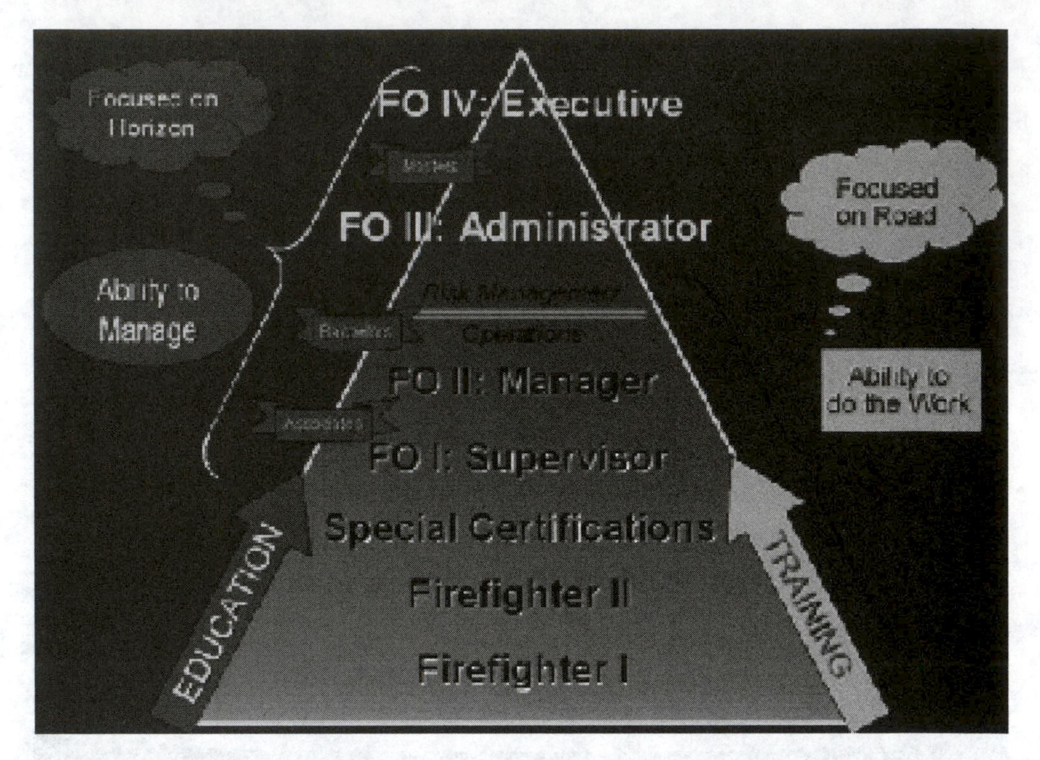

FIGURE 11.7 ◆ The pyramid illustrates the progression of education and training in the fires service as noted by the United States Fire Administration.

degree programs, and the state organizations representing fire chiefs, firefighters, volunteers, instructors, and other vital constituencies must be leaders. At the federal level, the USFA can bring the national stakeholders together to build momentum for this effort. An effective model for a state professional development "summit" was presented at the FESHE IV conference; it provides a plan of action for Washington State, including stakeholder involvement and consensus strategies. What should be the elements of a state professional development plan? In addition to spelling out who should be responsible for learning at each level of certification, it would recommend:

- The extent to which certifications should be granted academic credit
- The extent to which academic credit should be accepted toward satisfying standards
- The numbers and types of fire-related and general education courses
- The types of degrees—Associate of Arts/Associate of Science—transferable to baccalaureate programs versus terminal or nontransferable degrees

Only state and local leaders can make this happen. Contact the fire and emergency services leaders in your state, and urge these individuals to begin the difficult path of transforming this country's fire service's professional development into a national system.

Check the USFA/NFA Higher Education Web page for updates to this report. For questions about USFA's Higher Education programs, consult the website at *http://www.usfa.fema.gov/fire-service/nfa/higher-ed/he.shtm.*

TRADE

In 1984, state and local fire training systems were having trouble disseminating quality training programs effectively. This resulted in the establishment of TRADE. The Training

Resources and Data Exchange (TRADE) program is a regionally based network designed to foster the exchange of fire-related training information and resources among federal, state, and local government. Ten regional networks correspond to the existing federal regional boundaries that comprise the TRADE system. These networks provide a mechanism for the exchange of resources and materials within and among regions. Co-chairs serve as the points of contact for both intra-regional and inter-regional networking activities. One regional co-chair is selected from each of the state fire training systems, and the other is from the metropolitan fire services in each region. The TRADE network consists of the directors of the 50 state fire service training systems and senior executive training officers from the largest fire departments in each state. The representatives are from metropolitan fire departments or a fire department that protects populations greater than 200,000 and/or have more than 400 uniformed personnel.

TRADE is a resource that offers instructors the opportunity to share resources with other instructors either through a Listserv or their website. This is not a typical Listserv, and there are limitations as to who can belong to TRADE; however, you can still take advantage of many of the resources that are available. The Listserv comes out once a week with a digest of questions to which other instructors are seeking answers.

The objectives of TRADE are to:

- Identify fire, rescue, and emergency medical services training needs at the regional level.
- Identify and exchange training programs and resources within regions and, whenever possible, replicate those resources.
- Provide to NFA an annual assessment of fire training resource needs within the region, together with recommendations as to how TRADE can better support federal, state, and local fire training systems.
- Identify national trends that impact on fire-related training and education.

Every two years all participating members are invited to attend a national TRADE conference held at NFA that provides structured opportunities for the exchange of fire-related training and educational materials as well as peer networking. On alternate years there is a meeting of the 20 regional TRADE cochairs. Each regional network meets periodically with its membership for the same purposes.

Activity

Construct a plan for your professional development as a fire service instructor. Include your goals for education. List the organizations you will join and what you expect to get out of the organization along with what your contribution to that organization will be. Be sure to list time frames to meet each goal and how you plan to accomplish them.

Bibliography

National Highway Safety Transportation Administration. August 2002. *National guidelines for educating EMS instructors*.

United States Fire Administration, *www.usfa.gov*.

Thiel, A., Stern, J., Kimball, J., and Hankin, N. 2003. *Trends and hazards in firefighter training*. (No. USFA-TR-100.) Emmitsburg, MD: Federal Emergency Management Agency.

Appendix A

The following list of resources is not exhaustive, but it gives enough to get started. The information was current when book went to press; however, some information may have changed.

Department of Labor
This group provides various guidelines and standard in reference to employment. It has a variety of curricula that is available mostly at no cost.

Department of Energy
The Department of Energy has curricula relating to the transportation of hazardous materials and radioactive materials.

Centers for Disease Control
The CDC has materials for infectious disease (ID) curricula. It provides much of this material in text, PowerPoint, and Adobe Acrobat files as free noncopyrighted materials. It also has a number of e-mail listservers that provide access to updated reports and news items free of charge.

Occupational Safety and Health Administration
OSHA has a wealth of worker safety standards and guidelines available on its website, as well as curricula.

Federal Emergency Management Agency
FEMA has a variety of standards and guidelines on its website. It also has a wealth of curricula, including disaster management and health and wellness. You will also find technical reports to use in your class as real-life scenarios.

FIRE RESOURCES

NATIONAL ASSOCIATIONS AND ORGANIZATIONS

Congressional Fire Services Institute
900 Second St., NE, Ste. 303
Washington, DC 20002
202-371-1277
Fax: 202-682-FIRE
info@cfsi.org
www.cfsi.org

Federal Emergency Management Agency—
U.S. Fire Administration,
National Fire Academy,
Emergency Management Institute,
National Emergency Training School
16825 S. Seton Ave.
Emmitsburg, MD 21727
301-447-1000
Fax: 301-447-1497
www.usfa.fema.gov

Fire and Emergency Manufacturers and Services Association
P.O. Box 147
Lynnfield, MA 01940-0147
781-334-2771
Fax: 781-334-2771
info@femsa.org
www.femsa.org
Member benefits: quarterly newsletter, FEMSA News; education on management and marketing techniques; discount programs; seminars; insurance; warning labels and user information guide.

Fire Department Safety Officers Association
P.O. Box 149
Ashland, MA 01721-0149
508-881-3114
Fax: 508-881-1128
fdsoa@fdsoa.org
www.fdsoa.org

Member benefits: Monthly newsletter *Health and Safety;* Safety-Gram, a guide for safety officers; safety publications on the Website; seminars; online forum.

International Association of Dive Rescue Specialists
201 N. Link Lane
Fort Collins, CO 80524
800-423-7791
Fax: 970-482-1562
info@iadrs.org
www.iadrs.org
Member benefits: Shared resources; *Searchlines* magazine; 24-hour help line; IADRS patch and ID kit; discounts on Dive Rescue International courses; access to discussion board; preferred pricing on products from SAR Manager and software from Sea Wolff Diving software company.

International Association of Fire Chiefs (IAFC)
4025 Fair Ridge Dr., Ste. 300
Fairfax, VA 22033-2868
703-273-0911
Fax: 703-273-9363
www.iafc.org
Member benefits: *On Scene* newsletter; discounted conference rates; discounted fees for distance learning programs; membership in regional chapters.

International Association of Fire Fighters (IAFF)
1750 New York Ave. NW, 3rd Floor
Washington, DC 20006
202-737-8484
Fax: 202-737-8418
www.iaff.org
Member benefits: Determined by each local union

International Fire Service Training Association
c/o Oklahoma State University
930 N. Willis
Stillwater, OK 74078-8045
800-654-4055; 744-5723
Fax: 405-744-8204
Customer_service@mail.ifsta.org
www.ifsta.org
Member benefits: Invitation-only conference and the publication *Speaking of Fire.*

International Rescue and Emergency Care Association
P.O. Box 43100
Minneapolis, MN 55443
800-221-3435
rescue@ireca.org
www.ireca.org
Member benefits: Reduced rates at annual conference; quarterly newsletter, *Rescuer;* website link to bookstore with reduced rates on rescue and EMS titles.

International Society of Fire Service Instructors
P.O. Box 2320
Stafford, VA 22555
800-435-0005
Fax: 540-657-0154
info@isfsi.org
www.isfsi.org
Member benefits: *Instruct-O-Gram,* monthly training outline; monthly publication, *The Voice;* training resources; toolkit.

National Association of Emergency Vehicle Technicians
151 Lexington
Shirley, NY 11967
800-446-2388
Fax: 704-278-1062
president@naevt.org
www.naevt.org
Member benefits: Facilitates seminars and workshops; NAEVT training; official recognition of emergency vehicle technicians; monthly newsletter, *The Technician.*

National Association of Fire Equipment Distributors
104 S. Michigan Ave., Ste. 300
Chicago, IL 60603
312-263-9300
Fax: 312-263-8111
www.nafed.org
Member benefits: Check website.

National Fire Academy
U.S. Fire Administration
16825 S. Seton Ave.
Emmitsburg, MD 21727
301-447-1000
Fax: 301-447-1052
www.usfa.fema.gov

National Fire Protection Association
1 Batterymarch Park
Quincy, MA 02269
800-344-3555; 617-770-3000
Fax: 617-770-0700
www.nfpa.org
Member benefits: Right to vote on codes and standards; right to talk to staff for information and for interpretation of codes; 10% discount on any purchases; website benefits, including accessing certain electrical code information.

National Highway Traffic Safety Administration, EMS Division
400 7th St. SW
Washington, DC 20590
202-366-5440
Fax: 202-366-7721
www.nhtsa.dot.gov/people/injury/ems

National Institute for Urban Search and Rescue
P.O. Box 91648
Santa Barbara, CA 93190
805-569-5066
fax: 805-966-6178
niusr@cox.com
www.niusr.org

National Society of Executive Fire Officers
503-655-8542 (Ext. 204)
Fax: 252-399-2893
www.nsefo.org
Member benefits: Provides executive officer development focus through annual leadership conference; annual recognition of outstanding research projects; support of National Fire Academy and United States Fire Academy programs; quarterly newsletters.

National Volunteer Fire Council
1050 17th St. NW, Ste. 490
Washington, DC 20036
202-887-5700; 888-ASK-NVFC
Fax: 202-887-5291
www.nvfc.org/
NVFC publications, materials

U.S. Department of Transportation
400 7th St. SW
Washington, DC 20590
202-366-4000
Fax: 202-366-7342
training@rspa.dot.gov
dot.comments@ost.dot.gov
www.dot.gov

FIRE TRAINING BY STATE

California

California Department of Forestry & Fire Protection Academy
4501 State Highway 104
Ione, CA 95640-9705
209-274-2426
Fax: 209-274-2034
www.fire.ca.gov

California Fire Academy
836 Asilomar Blvd.
Pacific Grove, CA 93950
831-646-4240
Fax: 831-655-8400
nrodda@earth.npc.cc.ca.us

Connecticut

Wolcott Regional Fire Training School
29 Winterbrook Rd.
Wolcott, CT 06716
203-879-1559
Fax: 203-879-1559

Idaho

Idaho Emergency Services Training
650 W. State St., Room 324
Boise, ID 83720
208-334-3216
Fax: 208-334-2365

Iowa

Fire Services Training Bureau
3100 Fire Service Rd.
Ames, IA 50011-3100
888-469-2374; 515-294-6817
Fax: 515-294-2156

Kansas

Butler County Community College Fire Service Training Program
901 South Haverhill Rd.
El Dorado, KS 67042
316-321-2222
Fax: 316-322-3109

Dodge City Community College Fire Service Training Program
2501 North 14th Ave.
Dodge City, KS 67801
620-225-1321
Fax: 620-227-9319
info@dccc.cc.ks.us

Hutchinson Community College Fire Science Program
1300 North Plum
Hutchinson, KS 67501
620-662-3366 or 800-289-3501
Fax: 620-662-1090
www.hutchcc.edu/dept/6/fire

Kentucky

Northern Kentucky Technical College
90 Campbell Dr.
Highland Heights, KY 41076
859-441-2010
Fax: 859-441-4252

Michigan

Michigan Firefighters Training Council
7150 Harris Dr.
Lansing, MI 48913
517-322-5444
Fax: 517-322-6540

Montana

MSU Fire Services Training School
750 6th St. S. W., Ste. 205
Great Falls, MT 59406
406-761-7885, ext. 2;
800-294-5272, ext 2
Fax: 406-771-4317

New Jersey

Burlington County Emergency Services Training Center
53 Academy Dr.
Westhampton, NJ 08060
609-702-7157
Fax: 609-702-7100

Cumberland County 911 Emergency Communications & Training Center
637 Bridgeton Ave.
Bridgeton, NJ 08302
856-455-8526
Fax: 856-455-8515

Dempster Fire Service Training Center
350 Lawrence Station Rd.
Lawrenceville, NJ 08648
609-799-3245
Fax: 609-799-7087

Tennessee

Tennessee State Fire School
1303 Old Fort Pkwy.
Murfreesboro, TN 37129
615-898-8010
Fax: 615-893-4184

Texas

Collin County Community College Fire Science Dept.
2200 W. University
McKinney, TX 75071
972-548-6790

Texas Engineering Extension Services
301 Tarrow, 1st Floor
College Station, TX 77840-7896
979-845-7641

ACC Fire Training Academy
114 W. 2nd St.
Taylor, TX 76574
512-365-1911

Virginia

Community Hospital of Roanoke Valley College of Health Sciences, Fire & EMS Technology
920 South Jefferson St.
Roanoke, VA 24016
540-985-8273
Fax: 540-224-4404

Wisconsin

Hazmat Training Center Lakeshore Technical College
1290 North Ave.
Cleveland, WI 53015
888-468-6582

Milwaukee Area Technical College
700 W State St.
Milwaukee, WI 53233-1433
414-297-6370
www.matc.edu/utility/clas/prog/prot/fire.htm

Nicolet Area Technical College
P.O. Box 518
Rhinelander, WI 54501
715-365-4410, 800-544-3039
Fax: 715-365-4445

FIRE AND RESCUE TRAINING BY STATE

Alabama

Alabama Fire College Personnel Standards Commission
2501 Phoenix Dr.
Tuscaloosa, AL 35405
205-391-3741

Alaska

Alaska State Department of Public Safety
Public Safety Fire Service Training
5700 E. Tudor Rd.
Anchorage, AK 99507
907-269-5789
Fax: 907-338-4375
mark_barker@ak.dps.state.ak
www.dps.state.ak.us/fire/asp/

Arizona

Arizona Department of Building and Fire Safety State Fire Marshal's Office (Maricopa County)
1110 W. Washington, Ste. 100
Phoenix, AZ 85007
602-364-1003
Fax: 602-364-1052

Arizona Department of Building & Fire Safety State Fire Marshal's Office (Pima County)
400 W. Congress, Ste. 121
Tucson, AZ 85701
520-628-6220
Fax: 520-628-6930

Arkansas

Arkansas Fire Academy
P.O. Box 3499
Camden, AR 71711
870-574-1521

Arkansas Fire Academy
Fire Training Satellite, Northeast Arkansas
3105 Fire Academy Dr.
Jonesboro, AR 72404
870-932-9703
Fax: 870-932-5944

Arkansas Fire Academy
Fire Training Satellite, Northwest Arkansas
118 Industrial
Lincoln, AR 72744
501-824-4045
Fax: 501-824-4540

Arkansas Fire Academy
Fire Training Satellite, North Central Arkansas
P.O. Box 818
Marshall, AR 72650
870-448-2030
Fax: 870-448-3099

Black River Technical College Fire Training Center
P.O. Box 468
Pocahontas, AR 72455
870-886-5750
Fax: 870-886-7481

California

California Department of Forestry & Fire Protection (CDF)
4501 Highway 104
Ione, CA 95640
209-274-2426
www.fire.ca.gov/FireMarshal/SFMtraining/trainingschedule.asp

Colorado

Colorado Division of Fire Safety Fire & Hazmat Training
700 Kipling St., Ste. 1000
Denver, CO 80215
303-239-4463
Fax: 303-239-4405

Connecticut

Commission on Fire Prevention & Control
Connecticut Fire Academy
34 Perimeter Rd.
Windsor Locks, CT 06096-1069
800-627-6363
Fax: 860-654-1889

Fairfield Fire Department Training Division
205 One Rod Hwy.
Fairfield, CT 06824
203-254-4709
Fax: 203-254-4719

Hartford County Regional Fire School
370 Spruce Brook Rd.
Berlin, CT 06037
860-828-3242
Fax: 860-828-9538

New Haven Fire Academy
230 Ella Grasso Blvd.
New Haven, CT 06516
203-946-6215
Fax: 203-946-7881

Valley Fire Chiefs' Training School
126 David Humphreys Rd.
Derby, CT 06488
203-736-6222
Fax: 203-736-6222

Delaware

Delaware State Fire School
1461 Chestnut Grove Rd.
Dover, DE 19904
302-739-4773
Fax: 302-739-6245
www.delawarestatefireschool.com

Florida

Florida State Fire College
11655 NW Gainesville Rd.
Ocala, FL 34482
352-369-2800
http://www.fldfs.com/sfm/

Georgia

Atlanta Fire Academy
407 Ashwood Ave., SW
Atlanta, GA 30315
404-624-0650
Fax: 404-624-0657

Hawaii

Hawaii State Fire Council
3375 Koapaka St., Ste. H-425
Honolulu, HI 96819
808-626-1589

Honolulu Fire Department Training Center
890 Valkenburgh St.
Honolulu, HI 96818
808-422-0827
Fax: 808-422-9691

Illinois

Illinois Fire Service Training Institute
University of Illinois
11 Gerty Dr.
Champaign, IL 61820
217-333-3800
iuforum@uiuc.edu
www.uiuc.edu

Kansas

Barton County Community College Fire Service Training Program
245 NE 30th Rd.
Great Bend, KS 67530
800-748-7594

Kansas State University
2610 Claflin Rd.
Manhattan, KS 66506
785-532-6011
www.ksu.edu

Labette County Community College
Fire Science Program and First Responder
200 South 14th St.
Parsons, KS 67357
620-421-6700 (Ext. 1271)
www.labette.cc.ks.us/

Kentucky

Kentucky State Fire-Rescue Training
Rowan Regional Training Center
609 Viking Dr.
Moorehead, KY 40351
606-784-1393

Kentucky Tech Fire-Rescue Training
385 Old Greensburg Rd.
Campbellsville, KY 42718
270-465-8633
Fax: 270-465-8730

Kentucky Tech Fire-Rescue Training
Central Office
2624 Research Park Dr.
Lexington, KY 40512-4092
800/782-6823
Fax: 859-246-3152

Kentucky Tech Fire-Rescue Training
1695 Shelter Lane
London, KY 40741
606-862-0318
Fax: 606-878-0288

Louisiana

Jefferson Parish Fire Training
200 East St.
Bridge City, LA 70094
504-436-9150
Fax: 504-436-9154

Maine

Maine Fire Training and Education
207-767-9555
Fax: 207-767-9678
www.mainefiretraining.net

Maryland

Annapolis City Fire Department
Training Division
1790 Forest Dr.
Anapolis, MD 21401
410-263-7975
Fax: 410-268-1846

Anne Arundel County Firefighter
Cadet Training Program
P.O. Box 276
Millersville, MD 21108
410-923-5646
Fax: 410-923-5646

Baltimore City Fire Academy
6720 Pulaski Hwy.
Baltimore, MD 21237
410-396-9984
Fax: 410-325-3456

Barnstable County Fire-Rescue
Training Academy
P.O. Box 746
Barnstable, MD 02630
508-771-5391
Fax: 508-790-3082

City of Salisbury Fire Department
Training Division
311 West Isabella St.
Salisbury, MD 21801
410-548-3134
Fax: 410-548-3121

Hagerstown City Fire Department
Training Division
25 West Church St.
Hagerstown, MD 21740
301-791-2544
Fax: 301-797-7448

Maryland Fire & Rescue Institute
University of Maryland
College Park, MD 20742
301-226-9960
Fax: 301-314-1497
Director@mfri.org
www.mfri.org

Massachusetts

Department of Fire Services
State Rd., P.O. Box 1025
Stow, MA 01775
978-567-3100

Mississippi

Mississippi State Fire Academy
1 Fire Academy USA
Jackson, MS 39208
601-932-2444
fireacademy@msfa.state.ms.us
www.doi.state.ms.us/fireacad/fa_home.htm

Missouri

Missouri Fire-Rescue Training Institute
University of Missouri-Columbia
240 Hinkle Building
Columbia, MO 65211
573-882-2121; 800-869-3476
www.mufrti.org

Montana

Silver Bow Fire Training Center
350 Josette Ave.
Butte, MT 59701
406-782-6090

Nebraska

Nebraska Fire Marshal Training Division
246 S. 14th St.
Lincoln, NE 68508
402-471-2027

Nevada

Clark County Fire Department
Training Center
4425 West Tropicana Ave.
Las Vegas, NV 89103
702-455-7700
Fax: 702-455-8349
www.co.clark.nv.us/fire/training.htm

Fire Science Academy/Crisis and
Emergency Management Institute
University of Nevada, Reno
P.O. Box 877
Carlin, NV 89822-0877
800-233-8928
fireacademy@unr.edu
www.fireacademy.unr.edu

Las Vegas Fire Department
Training Division
633 North Mojave Rd.
Las Vegas, NV 89101
702-229-0470
Fax: 702-388-2504

Nevada State Hazmat and Fire
Service Center
Fire Marshal's Office
107 Jacobsen Way
Carson City, NV 89711
775-687-4290
Fax: 775-687-5122

Truckee Meadows Community College
7000 Dandini Blvd.
Reno, NV 89512
775-673-7000
webmaster@tmcc.edu
www.tmcc.edu/

New Hampshire

New Hampshire Fire Academy
33 Hazen Dr.
Concord, NH 03305
603-271-2661

New Jersey

Atlantic County Fire Training
5033 English Creek Rd.
Egg Harbor Township, NJ 08234
609-407-6742
Fax: 609-407-6745

Bergen County Police/Fire Academy
281 Campgaw Rd.
Mahwah, NJ 07430
201-785-6040
Fax: 201-785-6030

Camden County Fire Academy
Lakeland Complex
Blackwood, NJ 08012
856-374-6167
Fax: 856-374-6218

Cape May County Fire Marshal
4 Moore Rd.
Cape May Court House, NJ 08210
609-465-2570

Middletown Township Fire Academy
P.O. Box 4074
Middletown, NJ 07748
732-671-7152

New Jersey Training and Certification
Department of Community Affairs
Division of Fire Safety
P.O. Box 809
Trenton, NJ 08625-0809
609-633-6070
Fax: 609-633-6134

Salem County Fire Training School
135 Cemetery Rd.
Woodstock, NJ 08098
856-769-3500
Fax: 856-769-3500
Fax: 856-769-3571

Union County Fire Academy
300 North Ave.
Westfield, NJ 07090
908-654-9881
Fax: 908-654-9851

Warren County Fire Academy
1024 Route 57
Washington, NJ 07882
908-835-2050

New York

New York Office of Fire Prevention
& Control
41 State St., 12th Floor
Albany, NY 12231-0001
518-474-6746

New York State Academy of Fire Science
600 College Ave.
Montour Falls, NY 14865
607-535-7136

North Carolina

North Carolina Fire-Rescue Commission
Office of State Fire Marshal
322 Chapanoke Road
Raleigh, NC 27603
919-661-5880

North Dakota

North Dakota Firefighter's Association
State Fire School
1641 Capitol Way
P.O. 6127
Bismarck, ND 58506-6127
701-222-2799
Fax: 701-222-2899

Ohio

Ohio Fire Academy
8895 E. Main St.
Reynoldsburg, OH 43068
614-752-7189
Fax: 614-752-7111

Oklahoma

Fire Service Training, Oklahoma University
1723 W. Tyler
Stillwater, OK 74078
800-304-5727; 405-744-5727
Fax: 405-744-7377

Oregon

Oregon Department of Public Safety
Standards and Training
550 N. Monmouth Ave.
Monmouth, OR 97361
503-378-2100 (Ext. 2233), to request a list of facilities

Pennsylvania

Pennsylvania State Fire Academy
1150 Riverside Dr.
Lewiston, PA 17044
717-248-1115 (Ext. 1979)
Fax: 717-248-3580

Rhode Island

State Fire Marshal's Office
Fire Education and Training
Coordination Board
24 Conway Ave.
North Kingstown, RI 02852
401-294-0861

South Carolina

Division of Fire and Life Safety
Fire Marshal's Office
South Carolina Fire Academy
141 Monticello Trail
Columbia, SC 29203
803-896-9800
www.llr.state.sc.us

South Dakota

South Dakota Fire Service Training
118 W. Capitol Ave.
Pierre, SD 57501
605-773-3562 (Al Christie)
605-874-8470 (Steve Harford)

Texas

ACC Fire Training Academy
114 W. 2nd St.
Taylor, TX 76574
512-365-1911

Austin Community College
Fire Training Academy
114 West 2nd St.
Taylor, Texas 76574
512-365-1911
Fax: 512-365-8619

Texas Engineering Extension Services
301 Tarrow, 1st Floor
College Station, TX 77840-7896
979-845-7641

Utah

Utah Fire & Rescue Academy State Fire
Services Training
3133 Mike Jense Parkway
Provo, UT 84601
888-548-7816 (toll free)

Vermont

Vermont Fire Service Training Council
Vermont Fire Academy
317 Sanatorium Rd., P.O. Box 53
Pittsford, VT 05763
800-615-3473
www.dps.state.vt.us/vfstc/council.html

Virginia

Fairfax County Fire and Rescue
Training Academy
4600 W. Ox Rd.
Fairfax, VA 22030
703-631-8121

Virginia Department of Fire Programs,
Area 1 Office
1500 E. Main St., 3rd Floor
Richmond, VA 23219
804-371-0280

Washington

Washington State Patrol Fire
Protection Bureau
Fire Training Academy
50810 SE Grouse Ridge Rd.
P.O. Box 1273 N Bend 98045
425-453-3000
www.wa.gov/wsp/fire/fireacad.htm#recruit

West Virginia

West Virginia University Fire Service
Extension
Monongahela Blvd.
P.O. 6610
Morgantown, WV 26506-6610
304-293-2106

Wisconsin

Chippewa Valley Technical College Fire
Training Program
620 West Claremont Ave.
Eau Claire, WI 54701
715-855-7500

Neeah Regional Fire Training Center Fox
Valley Technical College
1470 Tullar Rd.
Neeah, WI 54956
920-751-5050
Fax: 920-751-5058

Northcentral Technical College
1000 West Campus Dr.
Wausau, WI 54401
715-675-3331; 888-NTC-7411
www.northcentral.tec.wi.us

Northeast Wisconsin Technical College
Fire Training Center
P.O. Box 19042
Green Bay, WI 54307-9042
920-498-5603
www.nwtc.tec.wi.us/

Southwest Wisconsin Technical College
1800 Bronson Blvd.
Fennimore, WI 53809
800-362-3322

Western Wisconsin Technical College
304 6th St. N
LaCrosse, WI 54601
608-785-9200; 800-248-9982
Fax: 608-785-9289

Wisconsin Technical College System Board,
Fire Education and Training
310 Price Place, P.O. Box 7874
Madison, WI 53707
608-266-7289
Fax: 608-266-1690

Wyoming

Fire Prevention and Electrical Safety,
Training Division
Herschler Building-1, W
Cheyenne, WY 82002
307-777-7288
Fax: 307-777-7119

National

National Fire Training
National Fire Academy
16825 S. Seton Ave.
Emmitsburg, MD 21727
301-447-1000
Fax: 301-447-1052
www.usfa.fema.gov/nfa

Periodicals

Fire Rescue Magazine
http://www.jems.com/firerescue/

Fire Chief
http://www.firechief.com

Fire Engineering
http://fe.pennnet.com/home.cfm

Firehouse Magazine
http://www.firehouse.com/

Website Resources

Fire Rescue Village
http://www.firevillage.com/

Fire and Safety Group
http://www.fs-business.com/

NIOSH Firefighter Fatality Reports
http://www.cdc.gov/niosh/facerpts.html

Firefighter Close Call Stories
http://www.firefighterclosecalls.com/

Industrial Fire World
http://www.fireworld.com/

ISO Mitigation
http://www.isomitigation.com/

NFPA
http://www.nfpa.org

National Fire Service Incident Management System Consortium
http://www.ims-consortium.org/highway.htm

ADULT EDUCATION RESOURCES

Professional Organizations

American Association for Adult and Continuing Education (AAACE)
1200 19th Street NW, Suite 300
Washington, DC 20036-2401
202-429-5131
Fax: 202-223-4579

American Education Research Association
1230 17th Street NW
Washington, DC 20036-3078
202-223-9484

American Society for Training and Development (ASTD)
1640 King Street, Box 1443
Alexandria, VA 22313
800-NAT-ASTD (628-2783), Membership and information 703-683-8100, Customer Service; 703-683-8183, Information Center

International Association for Continuing Education; Training (IACET)
1101 Connecticut Avenue NW, Suite 300
Washington, DC 20036
202-857-1122

International Association of Facilitators
7630 W. 145th Street, Ste. 202
St. Paul, MN 55124
612-891-3541
Fax: 612-891-1800 fax

International Society for Technology in Education (ISTE)
1787 Agate Street
Eugene, OR 97403-1923
503-346-4414, Membership; 800-336-5191, Order Desk
503-346-2412, Distance Education
Fax: 503-346-5890

Meeting Planners International
1950 Stemmons Freeway, Suite 5018
Info Mart, Dallas, TX 75207-3109
214-746-5222 or 214-746-5272, Membership

National Association of Government Training and Development Directors (NAGTADD)
167 W. Main St., Ste. 600
Lexington, KY 40607
606-231-1948
Fax: 606-231-1928

Society for Applied Learning Technology (SALT)
50 Culpepper St.
Warrenton, VA 22186
800-457-6812 or 703-347-0055
Fax: 703-349-3169

Society for Human Resources Management (SHRM)
606 North Washington Street
Alexandria, VA 22314-1997
800-283-SHRM or 703-548-3440, Membership and information
703-548-6999 TDD
Fax: 703-836-0367

Toastmasters International
P.O. Box 9052
Mission Viejo, CA 92690
714-858-8255

PERIODICALS

Adult Education Quarterly
Adult Learning
American Association for Adult & Continuing Education

Presentations
Lakewood Publications, Inc.
Syllabus Magazine (n/c U.S., Canada; Mexico $24)
1307 S. Mary Avenue, Ste. 211
Sunnyvale, CA 94087
syllabus@netcom.com

Training Magazine
Lakewood Publications, Inc.
trainmag@aol.com

LEARNING THEORISTS

Malcolm Knowles—andragogy:
http://www.infed.org/thinkers/et-knowl.htm
http://www.newhorizons.org/future/Creating_the_Future/crfut_knowles.html

Alberta Bandura—social learning theory:
http://fates.cns.muskingum.edu/~psych/psycweb/history/bandura.htm
http://www.ship.edu/~cgboeree/bandura.html

Robert Gagne and his work:
http://www.psy.pdx.edu/PsiCafe/KeyTheorists/Gagne.htm
http://www.ittheory.com/gagne1.htm

Edward Thorndike—behavioral psychology, connectionism:
http://www.indiana.edu/%7Eintell/ethorndike.shtml
http://www.psy.pdx.edu/PsiCafe/KeyTheorists/Thorndike.htm

J. Bruner—constructivist:
http://www.infed.org/thinkers/bruner.htm
http://www.psy.pdx.edu/PsiCafe/KeyTheorists/Bruner.htm

BIBLIOGRAPHY

Fire Rescue Magazine (January 2004) 2004 Resource Guide.

Appendix B

National Fire Protection Association (NFPA) 1041 Job Performance Requirements (JPRs) for a Fire Service Instructor

The following are the Job Performance Requirements for a fire service instructor per NFPA 1041. The corresponding chapter of the text is cited at the end of each JPR.

INSTRUCTOR I

4-2.2 Assemble course materials given a specific topic so that the lesson plan, all materials, resources, and equipment needed to deliver the lesson are obtained. (Chapters 2, 6, 8)

4-2.3 Prepare training records and report forms, given policies and procedures and forms, so that required reports are accurately completed and submitted in accordance with the procedures. (Chapters 5, 6)

4-3.2 Review instructional materials, given the materials for a specific topic, target audience and learning environment, so that elements of the lesson plan, learning environment, and resources that need adaptation are identified. (Chapters 2, 6, 8)

4-3.3 Adapt a prepared lesson plan, given course materials and an assignment, so that the needs of the student and the objectives of the lesson plan are achieved. (Chapter 2)

4-4.2 Organize the classroom, laboratory, or outdoor learning environment, given a facility and an assignment, so that lighting, distractions, climate control or weather, noise control, seating, audiovisual equipment, teaching aids, and safety are considered. (Chapters 4, 6, 8)

4-4.3 Present prepared lessons, given a prepared lesson plan that specifies the presentation method(s), so that the method(s) indicated in the plan are used and the stated objectives or learning outcomes are achieved. (Chapter 8)

4-4.4 Adjust presentation, given a lesson plan and changing circumstances in the class environment, so that class continuity and the objectives or learning outcomes are achieved. (Chapter 2)

4-4.5 Adjust to differences in learning styles, abilities and behaviors, given the instructional environment, so that lesson objectives are accomplished, disruptive behavior is addressed and a safe learning environment is maintained. (Chapters 3, 4)

4-4.6 Operate audiovisual equipment, and demonstration devices, given a learning environment and equipment, so that the equipment functions properly. (Chapter 8)

4-4.7 Utilize audiovisual materials, given prepared topical media and equipment, so that the intended objectives are clearly presented, transitions between media and other parts of the presentation are smooth, and media is returned to storage. (Chapter 8)

4-5.2 Administer oral, written, and performance tests, given the lesson plan, evaluation instruments, and the evaluation procedures of the agency, so that the testing is conducted according to procedures and the security of the materials is maintained. (Chapter 7)

4-5.3 Grade student oral, written, or performance tests, given class answer sheets or skills checklists and appropriate answer keys, so the examinations are accurately graded and properly secured. (Chapter 7)

4-5.4 Report test results, given a set if test answer sheets or skills checklists, a report form policies and procedures for reporting, so that the results are accurately recorded, the forms are forwarded according to procedure, and unusual circumstances are reported. (Chapter 7)

4-5.5 Provide evaluation feedback to students, given evaluation data, so that the feedback is timely, specific enough for the student to make efforts to modify behavior, objective, clear, and relevant; include suggestions based on the data. (Chapters 1, 2)

INSTRUCTOR II

5-2.2 Schedule instructional sessions, given department scheduling policy, instructional resources, staff, facilities, and timeline for delivery, so that the specified sessions are delivered according to department policy. (Chapter 10)

5-2.3 Formulate budget needs, given training goals, agency budget policy, and current resources, so that the resources required to meet training goals are identified and documented. (Chapters 6, 10)

5-2.4 Acquire training resources, given an identified need, so that the resources are obtained within established timelines, budget constraints, and according to agency policy. (Chapters 10, 11)

5-2.5 Coordinate training record-keeping, given training forms, department policy, and training activity, so that all agency and legal requirements are met. (Chapters 5, 6, 10)

5-2.6 Evaluate instructors, given an evaluation form, department policy, and job performance requirements, so that the evaluation identifies areas of strengths and weaknesses, recommends changes in instructional style and communication methods, and provides opportunity for instructor feedback to the evaluator. (Chapters 7, 10)

5-3.2 Create a lesson plan, given a topic, audience characteristics, and a standard lesson plan format, so that the job performance requirements for the topic are achieved, and the plan includes learning objectives, a lesson outline, course materials, instructional aids, and an evaluation plan. (Chapters 2, 9)

5-3.3 Modify an existing lesson plan, given a topic, audience characteristics, and a lesson plan, so that the job performance requirements for the topic are achieved, and the plan includes learning objectives, a lesson outline, course materials, instructional aids, and an evaluation plan. (Chapters 2, 8, 9)

5-4.2 Conduct a class using a lesson plan that the instructor has prepared and that involves the utilization of multiple teaching methods and techniques, given a topic and a target audience, so that the lesson objectives are achieved. (Chapters 2, 8)

5-4.3 Supervise other instructors and students during high hazard training, given a training scenario with increased hazard exposure, so that applicable safety standards and practices are followed, and instructional goals are met. (Chapter 6)

5-5.2 Develop student evaluation instruments, given learning objectives, audience characteristics, and training goals, so that the evaluation instrument determines if the student has achieved the learning objectives, the instrument evaluates performance in an objective, reliable, and verifiable manner, and the evaluation instrument is bias-free to any audience or group. (Chapter 7)

5-5.3 Develop a class evaluation instrument, given agency policy and evaluation goals, so that students have the ability to provide feedback to the instructor on instructional methods, communication techniques, learning environment, course content, and student materials. (Chapters 7, 9)

5-5.4 Analyze student evaluation instruments, given test data, objectives and agency policies, so that validity is determined and necessary changes are accomplished. (Chapter 7)

INSTRUCTOR III

6-2.2 Administer a training record system, given agency policy and type of training activity to be documented, so that the information captured is concise, meets all agency and legal requirements, and can be readily accessed. (Chapters 5, 6, 10)

6-2.3 Develop recommendations for policies to support the training program, given agency policies and procedures and the training program goals, so that the training and agency goals are achieved. (Chapters 5, 10)

6-2.4 Select instructional staff, given personnel qualifications, instructional requirements, and agency policies and procedures, so that staff selection meets agency policies and achievement of agency and instructional goals. (Chapters 10, 11)

6-2.5 Construct a performance based instructor evaluation plan, given agency policies and procedures and job requirements, so that instructors are evaluated at regular intervals, following agency policies. (Chapter 7)

6-2.6 Write equipment purchasing specifications, given curriculum information, training goals, and agency guidelines, so that the equipment is appropriate and supports the curriculum. (Chapters 8, 10)

6-2.7 Present evaluation findings, conclusions, and recommendations to agency administrator, given data summaries and target audience, so that recommendations are unbiased, supported, and reflect agency goals, policies, and procedures. (Chapters 7, 10)

6-3.2 Conduct an agency needs analysis, given agency goals, so that instructional needs are identified. (Chapter 9)

6-3.3 Design programs or curriculums, given needs analysis and agency goals, so that the agency goals are supported, the knowledge and skills are job related, the design is performance based, adult learning principles are utilized, and the program meets time and budget constraints. (Chapter 9)

6-3.4 Modify an existing curriculum, given the curriculum, audience characteristics, learning objectives, instructional resources, and agency requirements, so that the curriculum meets the requirements of the agency, and the learning objectives are achieved. (Chapter 2)

6-3.5 Write program and course goals, given job performance requirements and needs analysis information, so that the goals are clear, concise, measurable, and correlate to agency goals. (Chapter 9)

6-3.6 Write course objectives, given JPRs, so that objectives are clear, concise measurable, and reflect specific tasks. (Chapter 9)

6-3.7 Construct a course content outline, given course objectives, reference sources, functional groupings and the agency structure, so that the content supports the agency structure and reflects current acceptable practices. (Chapter 9)

6-5.2 Develop a system for the acquisition, storage, and dissemination of evaluation results, given agency goals and policies, so that the goals are supported and those impacted by the information receive feedback consistent with agency policies, federal, state, and local laws. (Chapters 7, 10)

6-5.3 Develop course evaluation plan, given course objectives and agency policies, so that objectives are measured and agency policies are followed. (Chapters 7, 9, 10)

6-5.4 Create a program evaluation plan, given agency policies and procedures, so that instructors, course components, and facilities are evaluated and student input is obtained for course improvement. (Chapters 7, 9, 10)

Index

Page numbers followed by f indicate figure.

E

F

G

M

N